GEOLOGY AND MAN

TITLES OF RELATED INTEREST

GEOLOGY AND MAN

An introduction to applied earth science

Janet Watson FRS

Department of Geology,
Imperial College of Science and Technology,
University of London

London
GEORGE ALLEN & UNWIN
Boston   Sydney

George Allen & Unwin (Publishers) Ltd,
40 Museum Street, London WC1A 1LU, UK

George Allen & Unwin (Publishers) Ltd,
Park Lane, Hemel Hempstead, Herts HP2 4TE, UK

Allen & Unwin Inc.,
9 Winchester Terrace, Winchester, Mass. 01890, USA

George Allen & Unwin Australia Pty Ltd,
8 Napier Street, North Sydney, NSW 2060, Australia

First published in 1983

British Library Cataloguing in Publication Data

Watson, Janet
  Geology and man.
1. Earth sciences
I. Title
550   QE26.2
ISBN 0-04-553001-7
ISBN 0-04-553002-5 Pbk

Library of Congress Cataloging in Publication Data

Watson, Janet, 1923–
  Geology and man.
Includes bibliographical references and index.
1. Geology. I. Title.
QE26.2.W37 1983   550   83-2614
ISBN 0-04-553001-7
ISBN 0-04-553002-5 (pbk.)

Illustrated by Kidlington Graphics, Oxford

Set in 10 on 12 point Melior by
Computape (Pickering) Ltd, North Yorkshire
and printed in Great Britain
by Richard Clay (The Chaucer Press) Ltd,
Bungay, Suffolk

Preface

Many of the natural resources used by man come directly from the Earth and have been formed by the geological processes that are responsible for the evolution of the Earth as a whole. To understand the distribution of these resources, one must have a knowledge of their origins and geological significance. The Earth sciences therefore have a wide range of practical applications, extending into the fields of water supply, energy resources, mining, civil engineering, agriculture and public health. Most professional geologists have become involved at one time or another in problems relating to one or more applications of geology, and the need for expertise in these fields is likely to increase in the future.

Almost all students who graduate in Earth sciences take courses in applied geology. At an advanced level, they are well served by specialist textbooks dealing with individual subjects such as engineering geology or mining geology. It is, however, more difficult for the first-year student, who has not yet begun to specialise, to obtain a view of the whole field over which geological knowledge contributes to the solution of practical human problems. My aim in writing this book has been to provide a comprehensive introduction to the applications of the Earth sciences that can be read by anybody who has an understanding of general geological principles. The book is intended primarily for students in the first and second years of a degree course, but I hope that it may also interest A-level students and those people who come up against geological problems in the course of their work in neighbouring fields, such as environmental science.

The core of the book consists of six chapters dealing in turn with the main applications of geology. The first and last chapters set the scene and discuss the framework within which applied geologists operate. The penultimate chapter outlines the principal techniques used by applied geologists. In the interests of readability I have segregated a great deal of factual information in the tables and figures: these are essential parts of the book and need to be carefully digested.

I must end this preface by thanking the many geologists who have patiently answered my questions and punctured my illusions during the period over which this book has been written: in particular, I would like to thank my colleagues and students at Imperial College, from whom I have learned the fascination of applied geology.

Janet Watson
Imperial College, November 1982

Acknowledgements

I am particularly indebted to Dr M. Hale, who gave me the data for Figure 8.3, to Mr M. Poulsford for his help in tracking down photographs in the Institute of Geological Sciences collections, to Dr E. Robinson and Mr M. Gray for their help with Figure 5.1, and to Mr F. W. Dunning and Dr P. Garrard for supplying the originals of Figures 2.2 and 4.10. Other photographs were kindly supplied by Aerofilms Ltd, Barnaby's Picture Library and H.M. Ordnance Survey and the Institute of Geological Sciences.

My thanks are also due to all those individuals and organisations who have allowed me to use published or unpublished materials in the preparation of the illustrations: the principal source of each figure is noted in its caption in full detail. Numbers in parentheses below are text figure numbers:

Council of the Institution of Geologists (1.3); United States Geological Survey (2.6 & 6.5); J. L. Knill and the Geological Society (2.7 & 6.10); P. E. Potter and the American Association of Petroleum Geologists (AAPG) (3.6); AAPG (3.7a); J. R. Kyle and Elsevier (3.7b); G. E. Murray and the AAPG (3.8); P. R. Vail and the AAPG (3.9); K. A. Albright, Chevron Overseas Petroleum Inc. and the AAPG (3.10); R. H. Kirk and the AAPG (3.13); Figures 4.4 & 11 reprinted from J. McM. Moore 1982. *Metallization associated with acid magmatism*. New York: Wiley. © 1982 John Wiley & Sons Ltd, by permission of the publishers and author; C. J. Dixon and Chapman & Hall (4.5); D. S. Carruthers and the Editor, *Economic Geology* (4.9); the Director, Institute of Geological Sciences (5.2); Penguin Books (5.4); D. Brunsden and the Royal Society (6.2b); S. E. Hollingworth and the Geological Society (6.3); G. A. Kiersch (6.9); T. S. West and the Royal Society (7.3); W. B. Wilkinson and the Royal Society (7.5); P. Little, M. H. Martin and Applied Science Publishers (7.6); M. H. P. Bott and the Geological Society (8.2).

Contents

CONTENTS

List of tables

1 The human context

1.1 Introduction

Human communities have had a working knowledge of the land in which they live and a familiarity with useful rocks and minerals since prehistoric times. For many centuries, indeed, progress in the practical aspects of Earth science kept well ahead of theoretical developments. The 19th and 20th centuries, however, saw a revolution in our understanding of the structure and history of the Earth as a whole, as well as a growing demand for solutions to practical problems connected with the management of the Earth's resources.

With the increase of world population and the spread of industrial development, the need to assess the resources available to mankind and to make rational decisions about future developments has become more urgent, and many people concerned with government, industry and land use need to understand the geological factors affecting such decisions. This book is intended to provide a general introduction to the applied Earth sciences not only for those who intend to become geologists or geophysicists but also for those to whom geology is ancillary to civil engineering, mining, agriculture or environmental science. It does not aim to provide a professional training, for which more specialised texts are available, and it is written on the assumption that the reader is familiar with geological principles and terminology, or has access to standard introductory textbooks such as those listed in the bibliography.

No applied science constitutes a self-contained body of knowledge. Besides drawing on the basic principles of the parent science, the applied scientist has to take account of factors that have little to do with the discipline in question. Suppose, for example, that two deposits containing a metal in short supply are discovered, one in a sparsely populated area with poor communications and the other in an old-established farming region; given that geological factors are roughly equal, the decision as to which deposit to develop will probably be made on the basis of economic, social and political considerations. On the other hand, the financial or strategic importance of a really large find creates enormous pressure for development even where costs are high in terms of investment, social upheaval or risk. The context in which decisions are taken is determined by the demand for and supply of the commodity in question and by the standard of living, life style and values of the society involved. Although these topics cannot be explored in depth in a book of this kind, it may be worth while to survey briefly some more general aspects of the field in which Earth scientists have to operate.

1.2 Resources

Inorganic raw materials derived from the Earth can be classified from the consumer's point of view in terms of the uses to which they are put. This classification (Table 1.1) provides the basis on which the descriptive sections of the book are arranged. It will be clear from the start that demand for any

Table 1.1 Human needs and resources (resources not derived from geological materials are in parentheses).

Resource	Derived from
water supply	surface waters, ground water
energy	fossil fuels, nuclear, hydroelectric and thermal power (solar and wind power, organic materials)
metals	metalliferous ore deposits
building materials	rock, gravel, sand, clay, etc. (organic and synthetic materials)
non-metallic raw materials	rock materials (plant and animal products)
food	(plants and animals) supported by soils, fertilisers

commodity must vary from time to time and from place to place. The requirements of early man were restricted to a relatively small range of materials for building, pottery, ornamentation, tools and weapons. The Industrial Revolution brought not only a widening of requirements but also technical advances in mining, metallurgy and recovery processes generally. In the 20th century, consumption of many commodities has been closely related to living standards. The record of energy consumption in the mid-1970s, for example, reveals an almost straight-line relationship with income per head of population (Fig. 1.1). Production of many natural substances has increased by at least an order of magnitude since 1900 and it must be assumed that broadly similar relationships will be seen in the future. Approximate production figures for coal, oil and certain metals are given in later chapters.

If adequate provision is to be made for the future, realistic forecasts of needs, and estimates of reserves, are required. Neither are easy to come by. The **prediction of demand** involves making assumptions about factors which have influenced consumption in the past – changes in population, rates of economic development, price levels and the like. These assumptions are frequently undermined by unpredicted events (Fig. 1.2) or by technological advances that modify an established pattern, as the development of the internal combustion engine transformed the demand for petroleum in the early years of the 20th century. It is, therefore, scarcely surprising that published forecasts vary widely.

The **assessment of reserves** – the problem with which Earth scientists are more directly concerned – is almost equally difficult. Sup-

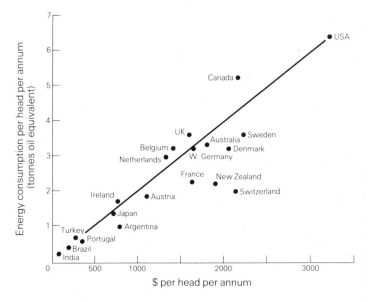

Figure 1.1 Energy consumption and income per head The relationships shown are based on figures compiled by British Petroleum for the mid-1970s, and have been modified as a result of a drastic rise in energy costs since 1974 (cf. Fig. 1.2).

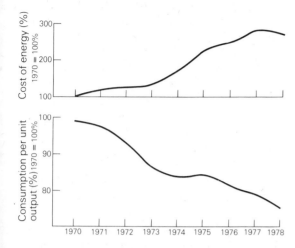

Figure 1.2 Energy efficiency The use of energy in relation to cost in the USA.

Table 1.2 'Published proved' oil reserves (in million barrels) based on statistical compilations of British Petroleum.

Year	North America	Middle East	World excluding USSR, China	Estimated world total
1949				~80 000
1969	48 300	332 800	480 600	540 600
1978	42 000	369 600	555 000	649 000

One barrel = 35 Imperial gallons or 42 US gallons.

plies of non-renewable resources such as coal or metalliferous ore deposits are finite. A single mine or group of mines passes through a characteristic cycle of discovery, development, production and decline (Fig. 1.3), and supplies are only maintained by the discovery of new deposits. Figures for reserves compiled at any time naturally take account only of deposits known at that time. Subsequent discoveries influence calculations made at a later date, with the rather paradoxical consequence that the known reserves of many commodities have increased over the years, despite the effects of consumption. 'Published proved reserves' of petroleum, for example, were at least eight times greater in 1978 than they were at the end of World War 2 as a result of the discovery of huge new oilfields (Table 1.2).

Figures for the known reserves of raw materials may also be modified by happenings which have nothing to do with geology. The outcome of international discussions concerning the legal position with respect to ownership of deep-sea mineral resources, for example, will radically affect estimates of future supplies of nickel, manganese and other metals. New methods of extraction and processing have on many occasions brought new types of deposits into the category of economic reserves – for example, the adoption of extraction techniques capable of dealing with copper ores averaging less than 1% Cu opened up the possibility of exploiting

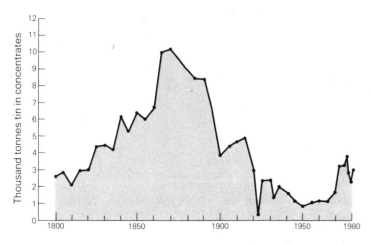

Figure 1.3 The history of a mining region Development, production and decline, illustrated by the production of tin in south-west England. The curve is unusual in that production on a small scale has continued intermittently for over two thousand years. At the peak of production in the 19th century over 300 producers were active. Only two were operating by 1960 and the future of tin mining in south-west England remains to be determined (based on Thorne, M. 1981. *British Geologist* **7**).

3

the enormous porphyry copper deposits which were too low grade to be handled by earlier techniques. Future advances may allow copper (now obtained mainly from sulphide minerals) to be extracted from common silicates, a process now made prohibitively expensive by its very high energy requirement. On a smaller scale, waste materials from 19th-century mine dumps have sometimes been reworked successfully for the original or a different metal. Current methods of extraction of hydrocarbons often leave over half the total volume in the ground and known hydrocarbon reserves could therefore be increased by improvement in recovery. An apparent rise in known reserves of a commodity may follow price increases which allow deposits formerly classified as uneconomic to be successfully exploited. This effect is illustrated in Table 1.3, which is based on a US Atomic Energy Commission assessment of uranium resources in the 1960s, and by the history of

aluminium production during World War 2 (Section 4.4.6).

The availability, as distinct from the existence, of resources is of course restricted by international political and social factors. The distribution of fossil fuels and metals is so uneven on a world scale that even large countries cannot be self-sufficient in all respects. The disparities between production and consumption of petroleum products (Fig. 1.4) illustrate this problem; they have profoundly influenced political developments in the Middle East during the past few decades. Among metals, large reserves of nickel, chromium and tin are confined to relatively few source areas.

Table 1.3 Exploitable reserves: variation with price illustrated by potential sources of uranium in USA, mid-1960s (based on figures of US Atomic Energy Commission).

Conventional resources
those which could be worked profitably at a price of $8–10 per pound (~$18–22 per kilo) of U_3O_8 in concentrate (1960s prices): in US, most conventional ore bodies are sandstone impregnations (Section 3.4.2).

Unconventional resources
those which could be worked for progressively higher prices as market values rose:

per pound U_3O_8 in concentrate (1960s prices)

$10–15	refractory mineral placers (monazite) marine phosphates uraniferous lignite, coal, oil
$15–30	rhyolitic tuffs carbonaceous shales ~65 parts per million (ppm) U
$30–50	acid or alkaline igneous rocks, ~15 ppm U, 45 ppm Th
$50–100	sea water ~3 parts per billion U
$100–500	all granites ~4–5 ppm U, 15 ppm Th

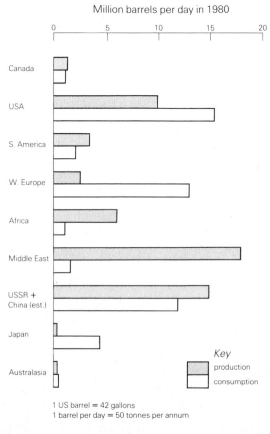

Million barrels per day in 1980

1 US barrel = 42 gallons
1 barrel per day = 50 tonnes per annum

Figure 1.4 Supply and demand Discrepancies illustrated by the production and consumption of oil in 1980 (based on a *Financial Times* compilation).

4

1.3 Disturbance of geological equilibria

The natural instability of the physical environment leads in the long term to changes in landforms, in sea level and in groundwater conditions. These changes have their impact on human activities – catastrophically in earthquakes, floods and volcanic eruptions; insidiously in the silting up of rivers, the erosion of cliffs and the rise or fall of sea level. Although little or nothing can be done to influence these geological processes in the long term, their incidental effects may be mitigated by the monitoring of potentially dangerous developments and by intelligent planning of mining, quarrying and engineering works. Problems of a rather similar kind arise where human intervention disturbs an existing state of equilibrium. Changes of

land use involving deforestation or the ploughing of grassland, for example, have been followed by erosion, flooding or deterioration of the soil; the construction of motorways or dams, by affecting the stability of natural slopes, can initiate phases of persistent landslipping (Fig. 1.5); the uncontrolled abstraction of water from wells can lead to the drawing in of salty or contaminated water to an underground reservoir.

The relationship between man and his geological environment raises problems which are the concern of the engineering geologist and environmental scientist. Although none of these problems is exclusively geological, all have geological aspects which are taken up in later chapters. An understanding of the factors that control geological processes in the surface parts of the Earth is essential if the effects of natural

Figure 1.5 Slope stability A rock slide resulting from the oversteepening of slopes during the construction of a highway in Colorado, 1957 (Barnaby).

changes are to be anticipated successfully and the unwanted side effects of human activities are to be minimised.

The adoption of an effective strategy to meet **geological hazards** is, however, seldom a matter for geologists alone. It depends, first, on the availability of the appropriate technical skills and financial resources and, secondly, on the decisions of governments and their electorates concerning the use to be made of these resources. In Britain, as in most industrial countries, activities involving large-scale disturbances of natural regimes by mining or civil engineering and release of potentially toxic substances are regulated by law. Although such legislation has proved extremely effective in the field of public health (in Britain, waterborne diseases have been almost eliminated since the mid-19th century, and mortality due to atmospheric pollution has been dramatically reduced in the present writer's lifetime), the direct costs are high. British industry, for example, spent an estimated £500 million annually in the mid-1970s on complying with regulations governing the emission of waste substances into the atmosphere. It is hardly surprising that the control of pollution is sometimes given a low priority in developing countries struggling to industrialise.

In mainland Europe, where the legislative framework varies from country to country, difficulties arise over the management of the great rivers that cross several frontiers. The potential of the Rhine as a source of water in the densely populated Netherlands is reduced by the presence of pollutants discharged into the river from industrial complexes upstream. Contamination of the Elbe by industrial effluents above and down stream of its point of entry into West Germany is demonstrated by the anomalous concentrations of heavy metals found in the river sediment (Fig. 1.6). International agreement on such problems remains for the future.

Even when financial and legal complications are set aside, it must be acknowledged that human actions have seldom been

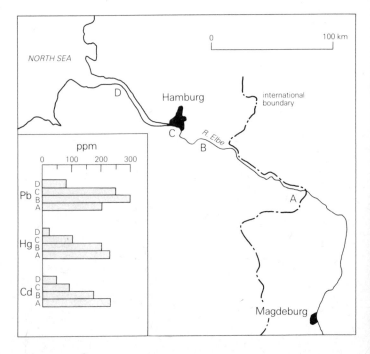

Figure 1.6 Pollution of an international waterway The heavy metal content of sediment in the River Elbe at four sites (based on an *Economist* compilation). In unpolluted areas which are geologically comparable, lead levels are usually <50 parts per million, mercury and cadmium much lower. High levels do occur, however, in volcanic and mineralised terrains.

decisively influenced by geological considerations. The rebuilding of cities such as San Francisco and Tokyo after destructive earthquakes, and the repopulation of the fertile slopes of volcanoes after major eruptions, show clearly enough the gap that exists between the geo-logic of a situation and the human response to that situation. The applied Earth scientist will never bridge this gap, but he may at least make sure that decisions are not taken in ignorance of geological realities.

References

Anderton, R., P. H. Bridges, M. R. Leeder and B. W. Sellwood 1979. *A dynamic stratigraphy of the British Isles*. London: George Allen & Unwin.

Greensmith, J. T. 1978. *Petrology of the sedimentary rocks*, 6th edn. London: George Allen & Unwin.

Hatch, F. H., A. K. Wells and M. K. Wells 1972. *Petrology of the igneous rocks*, 13th edn. London: George Allen & Unwin.

Holmes, A. 1965. *Principles of physical geology*, 2nd edn. London: Nelson.

Knill, J. L. (ed.) 1978. *Industrial geology*. Oxford: Oxford University Press.

Mason, R. 1978. *Petrology of the metamorphic rocks*. London: George Allen & Unwin.

Read, H. H. and J. Watson. *Introduction to geology*. 1961: vol. 1: *Principles*. 1975: vol. 2: *Earth history*. London: Macmillan.

Skinner, B. J. 1976. *Earth resources*, 2nd edn. Englewood Cliffs, NJ: Prentice-Hall.

Wyllie, P. J. 1976. *The way the Earth works*. New York: Wiley.

Note: geological and geophysical maps and memoirs covering the United Kingdom can be obtained from the Institute of Geological Sciences, Exhibition Rd, London SW7. Maps, specimens and displays illustrating many aspects of applied geology are on exhibition in the Geological Museum at the above address.

Open University publications listed in later bibliographies can be obtained from: The Open University, Walton Hall, Milton Keynes MK7 6AA, UK.

2 Water

2.1 Introduction

Water is essential for human survival as well as for the wellbeing of crops and livestock and for many industrial processes. The demand for water tends to rise steeply with living standards as industrial, domestic and recreational needs multiply; demand in Britain was of the order of 300 litres per head per day in the early 1970s. The availability of water is a limiting factor in the development of many semi-arid regions which could be made fertile by irrigation; and even where rainfall is moderate, densely populated communities such as those of Western Europe and North America have difficulty in maintaining adequate supplies of clean water. The problems involved in meeting the need for water vary according to the climate, topography and geology of an area and to the lifestyle of its population. All expedients depend on taking advantage of the natural circulation of water in the outer parts of the Earth and involve rerouting water from its original paths by modifications – usually minor but occasionally profound – to the natural system. The first sections of this chapter deal with factors controlling the general circulation of water and the later sections with water supplies for human use and with the management of these supplies in relation to the natural system.

2.1.1 The hydrological cycle
Water occurs in three settings of interest in the present context – in the solid **lithosphere** in the pores and larger spaces of rocks or in combination with rock-forming minerals; in the oceans, lakes, rivers and glaciers which collectively form the **hydrosphere**; and as vapour, clouds, snow and rain in the **atmosphere**. The mobility of water in all its phases allows it to migrate freely from one environment to another in response to temperature changes, gravitational forces and chemical and biological processes; the sum of all these migrations constitutes the **hydrological cycle** (Fig. 2.1).

Water enters the atmosphere as vapour released by evaporation and by the transpiration of plants and respiration of animals. It returns as rain or snow falling under gravity and collectively constituting the precipitation. A large portion of this precipitation re-enters the atmosphere more or less directly while the remainder, known collectively as the **runoff**, either joins the hydrosphere or seeps into the ground, thereby replenishing the surface waters and ground water respectively. Surface waters move under gravity towards the sea, repeated exchanges taking place both with ground water and with the atmosphere during migration. The sea receives water from rain and from rivers and loses it not only to the atmosphere but also to the lithosphere as **connate water** trapped in newly deposited sediments. The hydrological cycle is completed by the expulsion of connate waters during diagenesis. Accessions of **juvenile water** from the lower crust and mantle resulting from igneous activity and metamorphism are balanced to some extent by the transport of water to deep levels during subduction.

2.1.2 Water chemistry
As rain and snow normally contain few impurities other than dissolved CO_2, the runoff that feeds rivers and ground water is

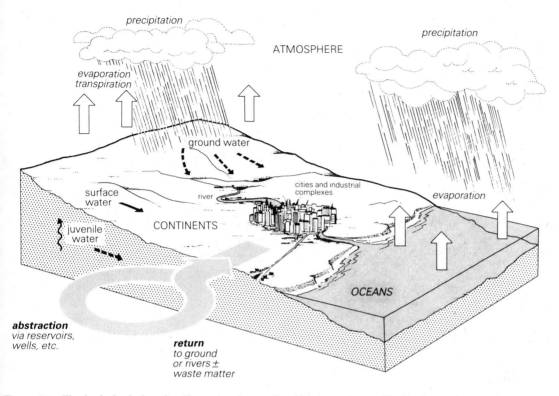

precipitation

ATMOSPHERE

precipitation

evaporation
transpiration

ground water

surface
water

river

cities and industrial
complexes

evaporation

juvenile
water

CONTINENTS

abstraction
via reservoirs,
wells, etc.

return
to ground
or rivers ±
waste matter

OCEANS

Figure 2.1   The hydrological cycle   Illustrating the way in which it may be modified by human intervention.

'fresh'. Both surface and ground waters, however, pick up dissolved and colloidal matter as a result of reactions with rock and soil, and receive organic materials released by biological (including human) activities. The total content of dissolved solids in surface waters is generally no more than a few tens or hundreds of parts per million, though it rises to much higher levels in regions of interior drainage in arid climates where salts become concentrated by evaporation. The composition of ground water varies with the character of the bedrock, the style of weathering, the **hydrogen ion content** (pH) which determines acidity (pH7 = neutral, <7 = acid, >7 = alkaline) and the **oxidation/reduction conditions** (Eh). Further local variations are induced by the influx of juvenile water carrying dissolved matter of magmatic origin or of waste matter from cities,

farms or factories. Table 2.1 gives data on the composition of waters in a variety of environments, together with a note on the acceptable limits for various components of water for human use.

The main inorganic components of surface and ground waters are cations of hydrogen, calcium, magnesium, sodium, potassium and the anions CO_3^{2-}, HCO_3^{1-} (carbonate and bicarbonate), Cl^{1-} (chloride), SO_4^{2-} (sulphate) and OH^{1-} (hydroxyl). **Hard water** has an excess of bicarbonate, resulting from solution of limestone, or other limy rocks, or of sulphate derived from gypsum from hydrogen sulphide. Reaction of hard water with soap gives an unpleasant scummy precipitate. Bicarbonate is said to produce **temporary** hardness because it can be removed by boiling which leads to precipitation of calcium carbonate. The **scale** that

9

Table 2.1 Water chemistry (this table summarises information used in Chs 2, 3, 4 and 7).

Total dissolved solids

typical soft water	20–60 ppm
hard water (Chalk of England)	up to 200 ppm
ground water in arid region	up to 4000 ppm
sea water	35 000 ppm (3.5%)
metalliferous brine (Red Sea deeps, Section 4.3.4)	~250 000 ppm
oilfield brine	up to 650 000 ppm

Permissible limits of total dissolved solids in Australian ground waters

domestic use	2000 ppm
irrigation	500–1000 ppm
cattle	10 000 ppm
sheep	15 000 ppm

Composition (total dissolved solids = 100%)

	HCO_3^-	Cl^-	SO_4^{2-}	Br^-	Na^+
river water	49.1	6.6	9.4	–	5.3
sea water	0.2	55.3	7.7	0.06	30.6

	K^+	Mg^{2+}	Ca^{2+}	Fe^{2+}	SiO_2
river water	1.9	3.4	12.6	0.5	11.0
sea water	1.1	3.7	1.2	–	–

Minor elements (ppm)
(1) Natural ground water and (2) ground water contaminated by land fill leachate, Long Island, New York (cf. Table 2.4): (3) water in geothermal area, Yellowstone National Park, USA; (4) Red Sea brine.

	(1)	*(2)*	*(3)*	*(4)*
arsenic	<11	0–29	>150	
boron	0.0	240–2500	—	8
copper	0.0	0–20	—	<1
iron	—		3–215	81
lead	4.0	0–6	—	<1
manganese			7–283	82
sulphur			up to 5000	
strontium		130–1400	—	48
zinc		10–80	—	5

clogs kettles, pipes and boilers in hardwater areas is a product of this reaction. Its accumulation is sometimes avoided by the use of water softeners which react with bicarbonate to give a soluble product. The hardness due to sulphate cannot be removed by boiling and is said to be **permanent**. A high content of sulphate makes water corrosive (owing to the formation of dilute sulphuric acid) and leads to damage to metallic structures, especially those of iron, and to concrete (Section 5.2.3). High levels of chloride make water undrinkable by humans and by farm stock. In many deep aquifers, such as that underlying the Great Artesian Basin of Australia, chloride levels may exceed safe limits for drinking and irrigation, while the brines of certain oilfields are much saltier than the sea (Table 2.1).

Organic matter is released to surface and ground waters as a result of bacterial and other decay processes as well as by the discharge of domestic, farm and industrial waste. Ammonia (CH_3) is among the commonest decay products and its abundance gives a rough and ready indication of organic pollution. The **biochemical oxygen demand** (BOD) measures the content of ammonia and other reducing organic compounds by reference to the amount of oxygen needed to oxidise them; in Britain, a BOD of 74 ppm is the recommended maximum for river waters containing effluent from sewage works. The principal health hazards that result from the use of water containing organic matter derived from sewage arise from the presence of the bacteria that transmit human diseases. These bacteria are destroyed along with the organic compounds by complete oxidation, but may persist in untreated water and in stagnant pools and streams.

The sea, which constitutes over 90% of all surface waters, carries some 3.5% of dissolved solids that include at least trace amounts of almost every known element (Table 2.1). Its salinity and composition remain constant, except in small enclosed basins where it may be modified by evaporation or by the influx of river water. The connate waters of marine sediments, and ground waters beneath marine basins, are generally saline, though they differ from sea water as a result of exchanges taking place during diagenesis. Saline and fresh ground waters merge near the coastline, the interface shifting seawards or landwards in response to natural or artificial changes in equilibrium.

Viewed in relation to the hydrological cycle as a whole, the communicating waters of the outer parts of the Earth maintain a rough chemical balance, in that gains of material resulting from solution or reaction in one environment are compensated by precipitation in others. Such changes are most important in the unsaturated zone above the water table and in the zone of diagenesis. Chemical changes in surface and ground waters are, naturally, complemented by modifications of the solid rock, which have practical consequences with respect to the development of soils (Section 7.2) and the redistribution of metals (Section 4.4.6).

2.1.3 Water for human use

Water supplies are drawn from fresh surface and ground waters which may be tapped close to the point where they are needed or transported for some distance via pipes, aqueducts, wells and boreholes; waste water may be returned directly to the region of abstraction or diverted via drains and sewers to another locality (Table 2.2). The principal considerations involved in the development of water resources tend to be those concerned with quality, reliability and cost. Where rainfall is adequate and the population sparse, water may be taken directly from rivers, lakes or shallow wells, and dirty water discharged again into the ground or the river system, with a minimum of disturbance to the natural system. The provision of supplies for cities commonly involves either the construction of storage reservoirs or the sinking of wells to a water-bearing bed or

Table 2.2 Sources of water supply.

From surface waters
springs, oases, waterholes – points of emergence of ground water
rivers, glaciers, lakes
aqueducts, canals, pipes – artificial conduits
storage reservoirs impounded by dams or contained in pits or quarries
saline waters via desalination plants
sewage works – water recycled after purification

From ground waters
surface deposits (alluvium, sand dunes, boulder clay, etc.) tapped by shallow wells
bedrock tapped by shallow wells
artesian basins tapped by deep wells
aquifers maintained by artificial recharge

aquifer at depth, or the recycling of waste water after expensive treatment procedures. Such procedures disturb the established equilibrium to a greater or lesser extent by setting up new base levels, by speeding or inhibiting the migration of ground water, lowering the water table, modifying the distribution of saline or contaminated ground waters, or releasing new solutes in effluent flows.

It is clear that problems concerning the provision of water supplies cannot be isolated from those involved in the disposal of dirty water and that the solutions adopted should be aimed at minimising the side effects of disturbances of the hydrological cycle. These questions will be illustrated by reference to a number of examples in Section 2.6; the disposal of waste water is dealt with in Chapter 7.

2.2 Atmosphere–hydrosphere relationships

The volume of the annual precipitation of rain and snow is controlled by climate and topography. Global variations can be related to the arrangement of climatic zones and more local variations mainly to the distribution of highland tracts with respect to the flow of moist air currents. Average annual figures expressed per unit area range from over 250 cm in some tropical latitudes to about 25 cm in arid zones.

Only a fraction of the annual rainfall becomes available as a potential source of water. Even in a temperate region such as Britain nearly half the precipitation is rapidly returned to the atmosphere by evaporation and by transpiration from the plant cover, whereas in hot or arid regions the loss is much higher. The Everglades region of Florida, for example, loses about 80% of the precipitation by evapotranspiration. The loss to be expected under given conditions from a continuous plant cover with access to adequate soil moisture defines the concept of the **potential evapotranspiration** (PE) which

Figure 2.2 A Welsh stream in spate Illustrating the increase in runoff after heavy rainfall (photo by F. W. Dunning).

increases with temperature, sunshine and windspeed and is inversely related to humidity. Long periods in which PE exceeds precipitation lead to the development of moisture deficits in the soil, while periods of excess precipitation give water surpluses. The **water balance** implied by these relationships varies both seasonally and over longer periods. Supply problems arise where water deficits persist for long periods; in semi-arid and arid regions they result from the combined effects of low rainfall and evaporation due to low atmospheric humidity and high daytime temperatures.

Water in excess of evapotranspiration provides the runoff which joins the surface and ground waters; according to climate, it ranges from less than 5 cm to over 250 cm per annum. Much of the surface water, including most of the products of short-lived storms (Fig. 2.2), follows an uninterrupted course to the sea. The limiting factor in the use of surface sources for water supply is often the **dry-weather flow** of a river – the flow to be expected after long periods of dry weather. Significant improvements can be achieved by construction of storage reservoirs in the upper parts of the catchment area which are used to regulate flow to the lower reaches.

In climates that allow a moderate proportion of the rainfall to join the runoff, all but the smallest streams are perennial and the drainage system discharges directly to the sea. In arid regions, many rivers are intermittent, though water commonly continues to flow in alluvium beneath the river bed and can be tapped in shallow excavations. Such rivers may have no outlet to the sea, feeding their occasional flood water to interior drainage basins where inland seas and ephemeral playa lakes are produced. The high evaporation in such basins leads to the concentration of salts in both surface and ground waters and locally to the deposition of evaporites, processes which may render shallow ground water unsuitable for human consumption or irrigation (Table 2.1). Much of the rainfall in hot semi-arid and arid zones falls in occasional heavy storms and is rapidly lost unless impounded by artificial means; farmers on the margins of deserts commonly build small earth dams to retain some of this intermittent supply.

2.3 Surface-water–groundwater relationships

Near the Earth's surface, water can percolate through the pore spaces and larger voids in rocks to produce an interconnected system of **ground water**. The **zone of saturation** in which all spaces are filled extends up to within a few metres or hundreds of metres of the surface. The **water table** is the upper surface of the saturated zone and its configuration and depth control the relationships of ground and surface waters. In broad terms, water soaks down through the unsaturated zone above the water table but seeps out of the saturated ground below the water table, a contrast of obvious importance in

relation to the siting of reservoirs and the disposal of waste (Fig. 2.3; see also Fig. 6.10 and Section 6.3.1).

The level of the water table is normally given by the level of water standing in open fractures or in wells and boreholes. The enclosing rock is saturated to a level up to a metre or so higher by water held by capillary forces in small pores. Above this **capillary fringe** conditions fluctuate; water movements are relatively rapid, reactions with rock-forming minerals take place and oxidising conditions usually prevail. Materials taken into solution in the unsaturated zone may be redeposited near the water table, giving cemented crusts or concentrations of metals derived from weathered ore deposits (Section 4.4.6).

The relief of the water table reflects that of the land surface in a subdued form (Fig. 2.3). Springs appear where the water table intersects the surface and swamps or lakes may mark areas fed by ground water from the saturated zone. In arid regions the water table lies well below the valley floors through most of the year, only locally inter-

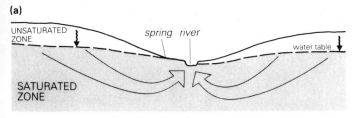

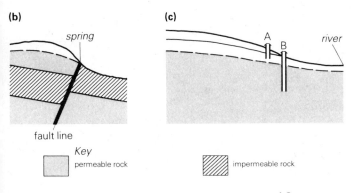

Figure 2.3 The water table (a) Forms of the water table and groundwater flow (arrows) in permeable rocks. (b) Influence of variations in permeability. (c) Effects of fluctuations in the water table in areas of low rainfall. The river runs dry when the water table stands at its lowest level; well A produces water only when the water table is high, whereas well B is perennial.

13

secting the surface to form **oases**. Rivers therefore lose water to the ground and flow only intermittently.

The ideal distribution of ground water outlined above is modified to a greater or lesser extent by variations in the permeability of the bedrock. Rocks with low permeability (Section 2.4) obstruct the flow of water in the unsaturated zone. Where they are interlayered with more permeable rocks, a **perched** water table may develop in the permeable beds, for example above clay layers in alluvium. Where more and less permeable rocks alternate at the land surface, water ponded against the upslope side of the less permeable rock may give rise to springs; such conditions may be produced at faulted junctions, at the outcrop of inclined strata or of dykes and other minor intrusions (Fig. 2.3). In the zone of saturation, a permeable aquifer may be effectively sealed off from communication with the general groundwater system by impermeable beds (**aquicludes**) above and below. The water in such a confined aquifer at depth is held under a hydrostatic head and should rise to the level of the water table at its unconfined edge when tapped by a well; a structure such as that illustrated in Figure 2.5 generates **artesian conditions** under which water may rise through wells to a piezometric level above the ground surface, thereby producing natural fountains.

2.4 Porosity and permeability

The capacity of a rock to store and transmit fluids, whether the fluid is water, brine, gas or oil, depends on two independent factors: the volume of pore spaces (porosity) in which fluid can be contained and the effectiveness with which the fluid can migrate through these spaces. The **porosity** is usually expressed as the ratio R = (volume of voids)/(volume of rock). It ranges from well over 50% in soils and unconsolidated sedi-

Table 2.3 Permeability and porosity.

	Porosity (%)	Permeability (cm s^{-1})
gravel	25–45	10^0–10^{-3}
fine sand and silt	45–50	10^{-3}–10^{-5}
clay and silty clay	35–55	10^{-5}–10^{-9}
sandstone	10–30	the reservoir rocks of many
limestone	5–25	oilfields fall in this range

ments or pyroclastics to almost nil in crystalline rocks at depth (Table 2.3). The **effective porosity** defined by reference to interconnected voids (i.e. setting aside sealed spaces such as vesicles or the chambers of foraminiferal tests) is of more interest to the water or oil geologist.

The voids in any rock are of two main types – the minute **pore spaces** between the constituent grains, and the larger but more widely separated **bedding planes**, **joints** and **faults**. Most of the processes of compaction and diagenesis taking place after deposition lead, by closer packing of the constituent grains and by the deposition of cement between grains, to decreases in pore space. The replacement of calcite by the denser mineral dolomite may, however, lead to increases in porosity in carbonate rocks; and previously lithified rocks that are exposed at the surface by erosion commonly develop open fractures as a result of the release of pressure. Weathering tends to increase porosity because it leads to disintegration of pre-existing fabrics and, especially in limestone terrains, to widening of fractures by solution. Thus several effects combine to produce a general decrease in porosity with depth.

The **permeability** of a rock is expressed by the speed with which a fluid flows through it along a hydraulic gradient and is, therefore, the principal factor determining its value in the supply of water, oil or gas. The coefficient of permeability or **hydraulic conductivity** is expressed in terms of Darcy's 'Law' which states that $K = Q/iA$ where K is the coefficient of permeability, Q the discharge, i ($i = \Delta h/l$) the hydraulic gradient and A the cross-sectional area of the sample.

Rocks that contain few discontinuities, and in which the interstitial pore spaces are small, have a low permeability, as capillary forces retard the migration of pore fluids. Clays, evaporites, many limestones and crystalline rocks at depth fall into these categories (Table 2.3). Coarser-grained sedimentary rocks such as sandstones and conglomerates, in which capillary forces have less effect, have a higher permeability, as have limestones made cavernous by solution or dolomitisation, and most weathered or strongly jointed rocks. The rate of transmission via intercommunicating fractures tends to be much greater than that through pore spaces, especially in limestones where discontinuities may be widened by solution; rocks in which flow takes place mainly via fractures are said to be **pervious**. The measurement of permeability in the laboratory, achieved by passing water through samples under known conditions, gives results which depend on transmission through the smaller pore spaces and which may bear little relation to the behaviour of the sampled rocks *in situ*. The measurement *in situ* of rate of fall and subsequent recovery of water levels in test boreholes sunk at known distances from a well from which water is being pumped, give more realistic values for hydraulic conductivity in that they take account of transmission along fractures and of local variations in permeability.

2.5 Groundwater flow

Ground water under a hydraulic head moves from regions of high potential to regions of low potential, draining (in broad terms) from highland regions towards lower ground. At shallow depths and in homogeneous rocks, the flow lines in plan are roughly perpendicular to the contours of the water table, the water descending beneath uplands and interfluves and ascending towards valleys and hollows which intersect the water table.

At somewhat deeper levels, complex migrations establish circulation cells, the boundaries of which correspond roughly with the lines of the main rivers. At considerable depths, high hydrostatic pressures tend to drive water upwards, so that descending meteoric waters may become mixed with connate waters expelled during compaction or with juvenile waters of magmatic origin.

The general circulation systems outlined above are always modified by the geological structure. Flow is impeded by rocks of low permeability and flow lines are diverted near the boundaries of impermeable units. Flow in a strongly jointed aquifer is guided by the orientation of the joints through which most of the water moves. In confined aquifers, flow takes place predominantly parallel to the surfaces of the aquifer, although there is usually some leakage into and through the adjacent aquicludes. Large-scale contrasts of permeability – for example, between a crystalline basement and its sedimentary cover – may determine the shape and size of the groundwater circulation cells. In accordance with Darcy's 'Law' (Section 2.4) the rate of flow of water in an aquifer depends on both the hydraulic head and the capacity of the aquifer to transmit fluid by any mechanism. Open fractures and gravels with large interconnected spaces provide conduits in which the mode and rate of flow differ little from those in a surface stream.

The direction and rate of groundwater migrations can be monitored by various means. Direct methods involve the introduction and tracing of a dye or other harmless marker substance. Indirect methods involve mathematical modelling of groundwater flow in relation to known points of recharge and abstraction. Since the early 1950s, extensive use has been made of the concentration of tritium (^{3}H) in rain water. Testing of nuclear devices in the 1950s and 1960s led to a sharp rise in tritium in surface- and subsequently in ground water. Tritium concentrations have been used to investigate the rate of recharge of aquifers, where recharge is due to the entry of rain water rather than to seepage of mixed ground water from adjacent strata. A study in southeastern South Australia, for example, showed that rain falling on a line of hills took 30–40 years to reach an aquifer beneath the plains 10–15 km away.

2.6 Management of water resources

The complexity of the relationships between surface and ground waters makes it inevitable that changes at one point should lead to adjustments elsewhere. Such adjustments take place naturally in response to a number of different causes. Short-term fluctuations in the water table, for example, follow variations in rainfall, whereas long-term changes in drainage follow a change in base level. The abstraction of water for human use is followed by similar adjustments and the management of water supplies therefore calls into play regional as well as local factors. The disturbance of the natural system involved in the supply of water to small towns and rural communities is often minor, since supplies are drawn mainly from rivers, lakes and shallow wells, and waste is returned to the same general area with little net loss. Modern urban communities with populations of a million or more, on the other hand, draw their supplies from distant sources, from deep aquifers, or from large reservoirs; their waste products may seriously contaminate the lower reaches of the drainage system or be discharged direct to the sea. These various alternatives are illustrated in the examples given below.

The importance of reservoirs for the storage of water and regulation of rivers has already been mentioned. It will be seen in later chapters that reservoirs serve other needs besides those of water supply. The siting and construction of dams capable of impounding a large volume of water, and problems connected with reservoir management, are dealt with in Section 6.3.

2.6.1 England and Wales

Consumption of water by the large, highly industrialised population of England and Wales doubled over the 30 years to 1970 when it stood at over 9000 million gallons per day. As the climate is temperate and rainfall moderate there is, in normal circumstances, no overall shortage of water, but there are major problems concerned with distribution – the bulk of the population is concentrated in southern and eastern England, while most of the rain falls in the highland areas of the west – and with the pollution of rivers below industrial centres.

At the present day, 70–75% of the water used in England and Wales comes from surface waters and 25–30% from ground water. Water supply and the treatment of dirty water in England and Wales are organised on a regional basis by ten regional water authorities (RWAs) whose boundaries coincide roughly with those of natural drainage basins (Fig. 2.4).

The RWAs were set up in 1973 under a co-ordinating National Water Council to replace a medley of local organisations, often dealing separately with water supply, river management and sewerage, which had grown up over many centuries. They were intended to provide a framework within which resources needed for the whole country could be integrated. For most purposes, the regions are self-sufficient, though transfers across boundaries to the drier regions are made on a relatively small scale and larger schemes by which water from the Severn basin could be fed into the headwaters of the Thames have been envisaged.

The supply system in England and Wales depends on the integration of surface and groundwater resources. In its natural condi-

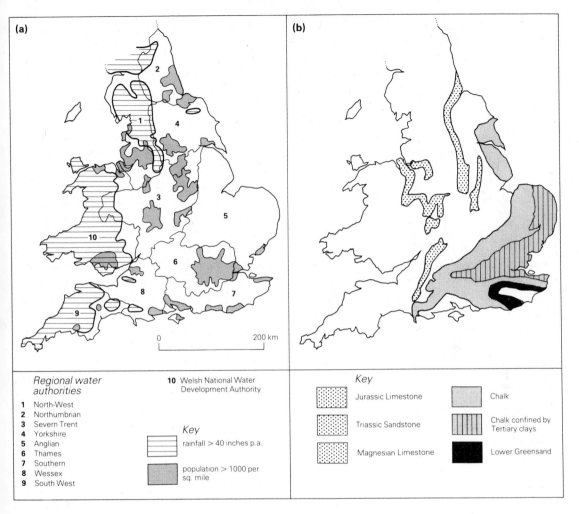

(a)

(b)

Regional water
authorities

1 North-West
2 Northumbrian
3 Severn Trent
4 Yorkshire
5 Anglian
6 Thames
7 Southern
8 Wessex
9 South West

10 Welsh National Water
Development Authority

Key

rainfall > 40 inches p.a.

population > 1000 per
sq. mile

Key

Jurassic Limestone

Triassic Sandstone

Magnesian Limestone

Chalk

Chalk confined by
Tertiary clays

Lower Greensand

0   200 km

Figure 2.4 Water supply in England and Wales   (a) Domains of the statutory regional water authorities in relation to areas of high rainfall and centres of population. (b) The principal aquifers.

tion, a river gives a secure water supply proportional to the dry-weather flow (Section 2.2), which may be no more than 10% of the average flow and which in some rivers includes more than 50% of treated effluent. The volume of many British rivers has been regulated by the construction of reservoirs near their headwaters which store water in times of plenty and feed it into the drainage system in times of drought. This system has the advantage of increasing the secure supply and of removing the need for expen-

sive and unsightly conduits by using the river itself for the transit of stored water. In the lower reaches of many rivers – especially in the Thames Valley around London – water is stored in further reservoirs at which it is purified and piped to the consumer. Interconnections with ground water in aquifers are made at many stages – in the London area, for example, water pumped from the Chalk is used to fill reservoirs in times of shortage and water pumped from surface sources is used to recharge the

17

aquifer when supplies permit. The principal disadvantage of the system lies in the need for large reservoirs both in upland areas of great natural beauty, such as Wales and the Pennines, and in lowland areas where land is in demand for agriculture and housing. The losses involved are offset to some extent by development of the reservoir sites as recreation areas available for sailing and fishing.

The lower reaches of rivers passing through thickly populated areas contain a high percentage of treated effluent. The quality of this recycled water is maintained by purification in sewage works along the river course. While bacterial contamination can be almost eliminated by standard procedures involving oxidation of organic waste, a variety of substances (nitrates from fertilisers, synthetic detergent, heavy metals used in industrial processes) are resistant to breakdown and, unless controlled at source, can pollute the discharge to the point of being unfit for drinking and, in extreme circumstances, toxic to plants, fish and

insects (Section 7.6). In 1973, most major British rivers were judged to contain potable water of the highest quality in their upper reaches, but the lower parts of the Trent and other rivers were too heavily polluted to support a satisfactory fauna and flora. A programme designed to improve the quality of Thames water was rewarded in the late 1970s by the reappearance of several species of river fish.

Underground water has traditionally been obtained from many shallow wells in alluvium and glacial deposits and from several major aquifers in the Mesozoic to Tertiary successions; most Palaeozoic strata in Britain are too tightly consolidated to hold large reserves.

The distribution, stratigraphical position and character of the principal aquifers are shown in Figure 2.4. Artesian conditions are provided in the London Basin, where Chalk and Greensand (Cretaceous) aquifers are confined in a broad syncline by clay aquicludes above and below (Fig. 2.5). The head of water sustained by the level of the water

Key

alluvium

Eocene aquiclude (London Clay)

Upper Cretaceous aquifer (Chalk)

Lower Cretaceous aquiclude (Gault) and underlying strata

→ freshwater flow

⇐ saline-water flow

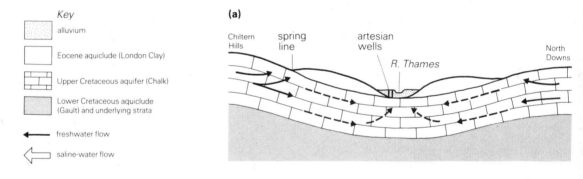

(a)

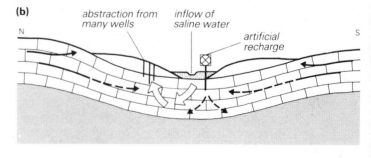

(b)

Figure 2.5 Water supply from the Chalk of the London basin (a) Natural conditions of groundwater flow in the unconfined (solid arrows) and confined (broken arrows) portions of the aquifer. (b) Modifications resulting from human intervention. For the sake of simplicity lower aquifers (Lower Cretaceous Greensands) are omitted. (After a report of the Water Resources Board.)

(a)

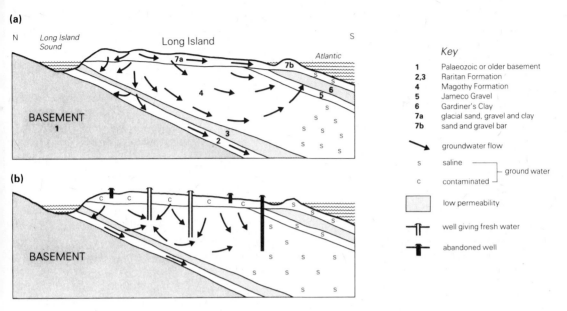

Figure 2.6 Groundwater conditions in Long Island, New York (a) Generalised conditions prior to development. (b) Present-day conditions. See also Table 2.4.

table in the hilly outcrop tracts of the Chalk in the North Downs and Chilterns was formerly sufficient to supply fountains in Trafalgar Square, but pressure has been reduced by excessive pumping and in 1965 the piezometric level was 80 m below mean sea level. This change in conditions has also led to inflow of salty water to the Chalk. It is balanced to some extent by artificial recharge of the aquifer (see above).

Artificial recharge plays a more significant role in maintaining supplies to Amsterdam, which for a century or more has drawn most of its water from sand dunes along the North Sea coast near Haarlem. In the thick sands of these dunes, a freshwater lens 40–90 m thick overlies saline ground waters. Extraction has exceeded natural recharge since 1915 and salt water has been drawn up into a brackish transition zone. Low-quality Rhine water carrying 5–10 mg l^{-1} of dissolved solids has been piped for a distance of over 50 km since 1957, and has effected considerable

Table 2.4 History of water supply in Long Island (see Fig. 2.6).

original state	in balance, saltwater front near south coast in glacial sediments, offshore in underlying aquifers: no groundwater pollution
early development	many shallow wells in glacial deposits, cesspools return waste water to same deposits, little net loss, saltwater front almost undisturbed
19th century	shallow wells partly contaminated and abandoned, public wells deepened to tap Jameco Gravel and Megothy Formation: cesspools return waste water to shallow levels, deeper aquifers depleted, salt water moves inshore in them
early decades 20th century	large public sewage systems constructed, effluent discharged to sea with net loss of fresh water, rapid encroachment of salt water in aquifers – western Long Island water supply piped from New York (c. 1950)

See Table 2.1 for analysis of polluted ground water.

19

changes of groundwater chemistry in the aquifer which it recharges (cf. Section 1.3).

2.6.2   New York and Long Island

Unlike most large European cities, New York stands on crystalline rocks and is backed by the Appalachian Mountains from which strong rivers flow towards the Atlantic coast. An initial period of reliance on water from shallow wells and small reservoirs was brought to an end both by the increase in demand and by repeated outbreaks of cholera and other waterborne diseases caused by contaminated supplies. Since the 1830s, supplies have been drawn mainly from the more distant rivers and, as the quality of the readily accessible Hudson River was quickly affected by industrial development along its course, reservoirs and aqueducts were constructed to bring supplies from the Catskill Mountains and the headwaters of the Delaware River some 200–300 km from the city. Severe shortages were felt in a series of dry years during the 1960s, when water from the Hudson River had to be purified to supplement the normal supply; future needs may be met by recycling water from this source.

Long Island stands on morainic and outwash sands and gravels underlain by a southward-dipping wedge of coastal sediments above the crystalline basement (Fig. 2.6). Cretaceous sandstone, Pleistocene gravel and superficial sands have all supplied water during the development of the area. The small size of the catchment area and the effects of increasing contamination have led to progressive changes in the pattern of exploitation, as illustrated in Figure 2.6 and Table 2.4. It will be seen that, over much of the area, contamination from individual cesspools put an early end to the use of water from superficial deposits, and that the subsequent development of public sewage works discharging direct to the sea increased the net loss of fresh water and was associated with an encroachment of salt water in the deeper aquifers. The more densely populated western part of the island can no longer be supplied entirely from underground sources and it receives piped water from New York.

2.6.3   Semi-arid and arid regions

Australia has a low average rainfall and, although 80% of the population is concentrated in the comparatively well watered coastal tracts, the arid interior regions support large numbers of sheep and cattle, as well as a scattered human population engaged in farming and mining. Supplies to the major eastern towns are piped, often for long distances, from storage reservoirs in the mountainous hinterland; the construction of these reservoirs demands a large capital outlay. The Great Artesian Basin in which the main aquifers are Mesozoic sandstones provides ground water from deep wells to a very large interior area. The value of this supply, like that of other aquifers in semi-desert areas, is limited by a high content of dissolved salts which may rise above the acceptable levels for irrigation and watering of stock (Table 2.1). A number of small desalination plants, mostly operating on salty ground waters, are now used to supplement natural supplies. Plans for the future envisage the integration of existing supply systems with thermal or nuclear desalination plants processing sea water for the cities.

Water supply problems are of particular importance in the arid Middle East where rapid expansion has accompanied the development of oilfields. Surface and shallow groundwater supplies, which tend to pick up salts from evaporitic sediments, are supplemented in some areas by water from deep aquifers. The Nubian Sandstone, a late Mesozoic fluviatile formation which underlies large areas in Libya, Egypt and Sinai, is a good aquifer. Studies in Sinai suggest, however, that much of the water in this aquifer dates from a pluvial period 30 000 years ago, a conclusion which raises the question of whether natural recharge under the present climatic conditions would replace water removed by pumping.

Tehran, the capital city of Iran, is supplied by an integrated system of surface and ground waters. Aqueducts bring water from the Karaj River some 40 km to the west of the city, where a large dam completed in the 1960s impounds a storage reservoir. Ground water is obtained from thick alluvial deposits which represent debris shed from the recently elevated Alborz Mountains north of the city (Fig. 2.7). The alluvium is poorly sorted and patchily cemented (the cement including gypsum in places), and porosities range upwards from about 24%. Grain size tends to diminish southwards and southwestwards from the mountain front and the many interleaved layers of fine sediment with lower permeability reduce the flow across the bedding. Groundwater flow, channelled by the bedding, is generally south-

wards from the mountain front where water is fed into the alluvium through the beds of a few permanent rivers and a number of intermittent streams, but is complicated by structural disturbances and variations in permeability. The lowest units (the A beds, presumed to be Pliocene), are rucked into east–west folds and water flowing in the overlying B, C and D beds (Pleistocene to Recent) is obstructed by ridges which represent partly eroded anticlines (Fig. 2.7). Surface waters emerging from the mountains are alkaline but contain less than 200 ppm dissolved solids. Ground waters below the city, on the other hand, range up to 1000 ppm dissolved solids, including chloride and sulphate derived from evaporitic minerals in the alluvium. Water in alluvium was traditionally extracted, as in many parts of the Middle East, by means of horizontal galleries (*qanats*) excavated in sloping surfaces of alluvium to intersect the water table. This system yields 22–24×10^6 m^3 per year, and

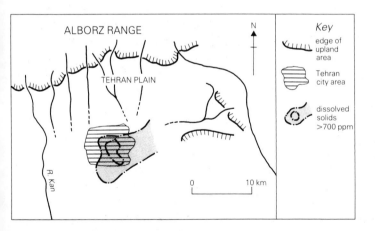

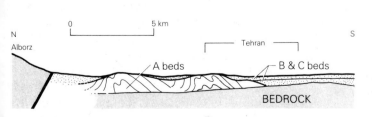

Figure 2.7 Water supply in arid regions Schematic map and section of the Tehran area showing the city in relation to intermittent rivers flowing out of the Alborz range and the relationships of water-bearing alluvial beds. In the shaded area, dissolved solids in ground water are usually >700 ppm (based on Knill, J. L. and K. S. Jones 1968. *Q. J. Engng Geol.* **1**).

pumping from wells and boreholes yields an additional volume of the same order. When these figures are set against the estimated natural recharge of the Tehran alluvium of $60–70 \times 10^6$ m^3 per year, it is clear that the present groundwater sources could yield very limited additional supplies in the future.

References

Blyth, F G. H. and M. de Freitas 1974. *A geology for engineers*. London: Edward Arnold.

Davis, S. N. and R. J. M. Dewiest 1966. *Hydrogeology*. New York: Wiley.

Garrels, R. M. and F. T. Mackenzie 1971. *Evolution of sedimentary rocks*. New York: Norton.

Open University 1974. *The Earth's physical resources*. Water resources, S226, Block 5. Milton Keynes: The Open University.

Open University 1975. *Water: origins and demand*. PT 272. Milton Keynes: The Open University.

3 Energy resources

3.1 Renewable and non-renewable resources

Prior to the Industrial Revolution, most of the world's energy requirements were met from sources which were, at least in principle, renewable, notably from wood and other plant and animal matter, from the power of water, wind and Sun, and from that of man and his beasts of burden. The energy needs of developing countries are met largely from the same sources today. In the long term, reliance on such sources has brought about profound changes in the natural equilibrium, in that the felling of trees for firewood has led to the deforestation of very large areas. The enormous increase in energy consumption during the past two centuries has been made possible by the exploitation of the fossil fuels – coal, oil and gas. Total consumption of these substances (which are used not only as fuels but also as feedstock for the chemical industry – see Ch. 5) increased threefold between 1900 and 1950 and had trebled again by about 1975. The fossil fuels, which now account for more than 90% of total energy supplies, represent a non-renewable resource, and while there is considerable uncertainty as to the extent of the reserves remaining, most authorities agree that at present rates of consumption they would be exhausted in, at best, a few hundred years. The correlation between standard of living and energy consumption (see Fig. 1.1) suggests that world demand is likely to increase (Fig. 3.1). These considerations, which imply that drastic changes in

the development of energy resources may be needed during the next half century, have stimulated research into both the development of nuclear fuels and the potentialities of renewable energy resources. Table 3.1 lists the principal resources now available or thought capable of development. Of these resources, solar power, wind power and wave power do not involve factors related to geology and will not be dealt with here.

3.2 Hydroelectric power

The energy of water flowing under gravity has been used for centuries to turn waterwheels and is now employed on a larger scale to drive turbines. Hydroelectric power produced in this way contributes 5–10% of the present production of energy. Water impounded in a storage reservoir on the course of a river is released to a generating station down stream, the head and volume of water controlling the output (Fig. 6.7). The potentiality of a scheme depends

Table 3.1 Sources of energy.

Renewable
solar power
water power: rivers, tides, waves
wind power
geothermal power
combustible organic materials

Non-renewable (fossil fuels)
solid fuels: the coal series, oil shales, oil sands
liquid fuels (oil)
natural gas
nuclear power: derived from fission or fusion processes

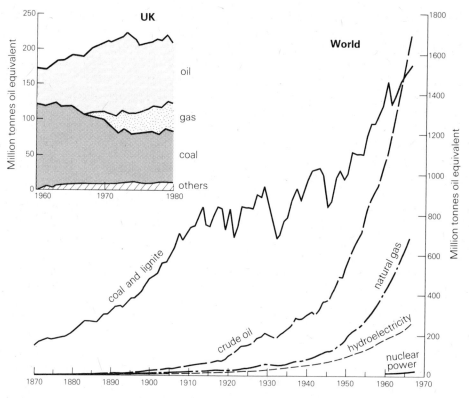

Figure 3.1 Growth in energy consumption Consumption during the century 1870–1970 (BP statistics) with, inset, the contributions of the principal fuels to energy production in the United Kingdom.

on rainfall, size of the catchment area, and availability of sites for dams and power stations. Most major hydroelectric schemes are located in or at the borders of upland areas with moderate to high annual rainfall, or near rapids or falls along the lower reaches of large rivers such as the Volta, Nile and Zambezi. The geological aspects of hydroelectric power schemes are principally those concerned with the construction of reservoirs and the siting of dams, which are examined in Section 6.3.

3.3 Geothermal energy

Heat flow from the interior of the Earth averages less than 0.1 W m^{-2} at the surface and provides a source of energy capable of development only in areas of anomalously high heat flow. The regional **geothermal gradient** varies from about 10 °C km^{-1} in old-established continental cratons to over 40–°C km^{-1} in regions of recent or active volcanicity. Higher figures are recorded locally both where circulating waters transfer heat from depth and in the immediate vicinity of igneous centres where geothermal gradients near the surface may be over 100 °C km^{-1}.

Geothermal energy is most often obtained from naturally occurring hot water or steam which can be used directly for space heating or the generation of electrical power. Corrosive solutes derived from juvenile (magmatic) water cause problems in some fields (cf. Table 2.1). 'Hot dry rock' systems involving the transfer of heat from dry rocks at depths

24

of a few kilometres are also technically feasible. The limitations of the resource are those imposed by geology, since the requisite conditions are widely attained only in active volcanic regions. Iceland obtains much of its energy, and North Island (New Zealand) and Italy 11 and 3% respectively, from geothermal sources; California, Central America, Chile, USSR and Indonesia make use of geothermal power, and many countries in the circum-Pacific volcanic zone or in the African rift-valley system may be expected to do so in future.

Iceland, which lacks fossil fuels, is exceptionally well placed to exploit geothermal power, since it lies athwart the mid-Atlantic ridge in a region of active sea-floor spreading. Eruption from fissures and central vents takes place intermittently within the 100 km tract which includes the ridge. Geothermal gradients reach about 100°C km^{-1} in this tract and decrease symmetrically away from it. The natural heat transfer in the ridge region is of the order of 55–60 MW km^{-2}. The pore fluids in the permeable lava pile consist largely of sea water and meteoric waters which have been heated by contact with the lavas to temperatures of 200–300°C at depths of a few kilometres. These circulating fluids are tapped near the surface and used to generate electricity; they are also piped directly to Reykjavik to provide space heating for factories and homes and for greenhouses where fruit, flowers and vegetables of species which seem incongruous in a northern land are cultivated.

Italy's geothermal power comes mainly from Lardarello in northern Italy where escaping steam was processed a few centuries ago for its content of boracic acid; there are no active volcanic centres other than fumaroles and it is assumed that the heat source responsible for the thermal anomaly is a young magmatic body at depth. Steam emerging at a temperature of around 200°C and pressure of 25 atmospheres contains a considerable proportion of juvenile fluid and is highly corrosive on account of its content of dissolved substances; it provided the first terrestrial sample of helium (previously identified only in the Sun). The corrosive character of the steam hindered development of an effective generating system before resistant materials for pipes and turbines came into use, and until recently it was employed, with considerable loss of efficiency, to heat pure water which in turn powered the generator.

Lower-grade geothermal heat sources exist well away from regions of active magmatism. In Britain, for example, two belts of relatively high heat flow have been identified and the potentiality of local anomalies within these belts is under investigation (Fig. 3.2). In south-west England, Hercynian granites rich in radioactive elements (K, U, Th) provide a local heat source which might be tapped in a hot dry rock system. Warm springs at Bath have made the town a health resort since Roman times, and in central-southern and eastern England hot water is encountered in, or at the margins of, Mesozoic–Tertiary sedimentary basins where heating may have been due partly to deep burial and partly to accessions of heat from the pre-Mesozoic basement. The Hampshire basin has many resemblances to the Paris basin which supplies energy for space heating in the suburbs of Paris. The long-term rate of renewal of heat in such areas may turn out to be below the economic rate of abstraction.

3.4 Nuclear fuels

3.4.1 The nuclear fuel cycle
The radioactive breakdown of uranium, thorium and other elements is a principal source of heat production in the Earth, and processes involving the emission of atomic particles have been harnessed to provide energy since World War 2. An explosive release of energy is employed in the detona-

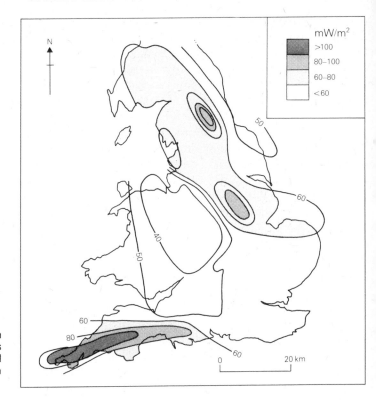

Figure 3.2 Heat flow Variations in England and Wales, indicating areas that might yield low-energy geothermal power. The heat flow is expressed in microwatts per square metre.

tion of nuclear bombs; controlled release in reactors generates power for industrial and domestic uses. About 200 commercial reactors are now in operation in 20 countries, mainly in Europe and North America; nuclear power supplied about 10% of the electricity used in Britain in 1976 and the development of nuclear power plants is regarded by many people as a means of filling the energy gap which will arise as fossil fuels are worked out. Such developments are, at the time of writing, the subject of public debate arising from the need to balance their undoubted economic and technical potentialities against possible dangers arising from the use of radioactive substances. The scale on which nuclear power finally comes to be employed and the technical systems which are adopted will depend largely on the solution of problems concerned with the possiblity of accidental leakage of radioactive materials, the development

of safe methods of disposal of radioactive waste products, the possibility of dissemination of radioactive materials by sabotage, or terrorist activity, and the control of nuclear weapons.

Two energy-producing reactions involve changes in atomic nuclei – **fission** of heavy nuclei and **fusion** of light nuclei. The nuclear fuel cycle employed today relies on the first and makes use of elements with high atomic numbers. Nuclear fusion takes place naturally in the Sun where hydrogen nuclei fuse to give helium and other light elements. A fusion process gives the motive power of the hydrogen bomb and may ultimately come to provide a source of power.

The principal raw material required for the generation of nuclear power by fission is uranium. The isotope uranium 235 is fissile; the more abundant uranium 238 is not, but can be converted into the short-lived fissile element plutonium 239. Reactors therefore make use either of uranium artificially enriched in 235 or of plutonium 239; **fast breeder reactors** utilise the conversion of uranium to plutonium. Fission is sustained by a chain reaction in which neutrons discarded by fission of one nucleus induce fission of another nucleus; a critical assembly is so balanced that each act of fission leads on average to one further fission. Plutonium in spent fuel elements can be recycled, but many other radioactive elements produced in reactors represent waste material for which safe repositories must be found (Section 7.6).

3.4.2 Sources of uranium

As the natural basis of the fission process, uranium has been the objective of intensive prospecting during the past few decades. Its high value allows rocks containing very low concentrations to be regarded as potential ore bodies, as was illustrated in Table 1.3. The principal ore minerals are uraninite and pitchblende (both $U^{4+}O_2$); others are coffinite (hydrated uranium silicate), together with other poorly defined compounds. Oxidised 'yellow products' (carnotite and uranophane) form characteristic stains around uranium concentrations. The principal means of prospecting depend on the radioactivity of the element. The **spectrometer**, which records variations in the emission of atomic particles on airborne or ground traverses, can be used, with certain reservations, to detect anomalous concentrations and to investigate anomalies found by routine geochemical reconnaissance (Ch. 8).

Because uranium, with its large atomic radius, is not readily accommodated in silicate lattices during the crystallisation of magmas, it tends to concentrate with other incompatible elements in the residual liquors. **Igneous rocks** derived from these residual magmas, such as alkali-granites, pegmatites and alkaline rocks, may therefore be enriched in both uranium and thorium and may carry uraninite, thorianite (ThO_2)

or monazite (Ce, La Y, Th) PO_4. Uranium deposits associated with granites in southwest England are mentioned in Section 4.3.

The bulk of the world's production comes, however, from **sedimentary rocks**, either from concentrations of detrital minerals or as precipitates from uranium-bearing groundwater. The key to the understanding of these deposits lies in the fact that uraninite and other minerals in which the element is tetravalent are readily decomposed under oxidising conditions. Uranium is therefore released during weathering to be redistributed by ground water and hydrothermal fluids.

Uranium deposits thought to have originated as placers are all of early Proterozoic ($\geqslant 2000$ Ma) age and some authorities consider that they were formed when low oxygen levels in the atmosphere allowed uraninite and pitchblende to survive weathering and concentrate along with other heavy minerals. The principal deposits of this kind are those in the Witwatersrand System of the Rand, South Africa which are dealt with elsewhere in connection with their gold content (in Section 4.4) and those of the Huronian Supergroup in the Canadian Shield. At Blind River (Ontario) uranium is concentrated in the thin pebbly basal bed of the Lorrain Quartzite which rests unconformably on a weathered surface of Archaean granite. The zone of mineralisation follows the unconformity over some hundreds of square kilometres as a yellow-stained layer no more than a metre or so thick which gives rise to pronounced anomalies on geochemical maps and spectrometer records.

Uranium carried in solution by circulating waters may be deposited in permeable sedimentary rocks at reduction sites where organic material is present or where the chemical environment changes. Over 90% of current production in the United States is based on deposits of these types, especially those of the continental red-bed successions

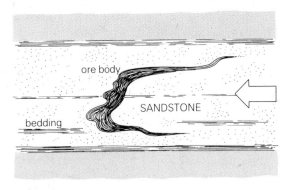

Figure 3.3 Uranium ore bodies Their common form in sandstone: the 'roll' is deposited by mineralising fluids advancing in the direction of the arrow. Such bodies are commonly up to a few metres in height.

of Mesozoic to Tertiary basins in Colorado, Utah and neighbouring states where deposition took place in permeable sandstones around plant fragments or as 'roll-front' forms at interfaces between pore fluids of differing composition (Fig. 3.3). In the arid terrains of western Australia and southwestern Africa, calcretes precipitated from saturated groundwaters (Section 7.2) are locally uraniferous where the bedrock contains unusually high levels of uranium. Sub-economic concentrations of uranium are not uncommon in carbonaceous sediments – lignite, oil shale, some coals and black shales – where the element may be associated with vanadium and/or copper (cf. Tables 1.3 & 4.8).

3.5 Fossil fuels: oil and gas

3.5.1 Introduction

The fuels which have been the basis of industrial development over the past two centuries (Fig. 3.1) are those derived from the organic debris in ancient sediments. Where complete oxidation is possible, organic matter is effectively destroyed by bacterial action in the top few metres of an accumulating pile of sediment. Where oxidation has not gone to completion, a residue of reduced organic carbon compounds remains in the rocks as hydrocarbons, carbohydrates and allied substances. Burning of these materials in the presence of air completes the oxidation process with the release of energy; hence their value as fuels.

The depositional environments and properties of the two main fossil fuels are distinct. Oil and natural gas are hydrocarbon mixtures derived mainly from marine microorganisms and are usually generated in marine basins. As they are fluids, they migrate readily through permeable rocks to accumulate in reservoirs which may be remote from the source rocks. The coal series is derived mainly from terrestrial or swamp vegetation, and coals are incorporated *in situ* in dominantly non-marine successions.

3.5.2 Geological environment

Most hydrocarbons in oil and gas fields are thought to have originated in fine-grained marine **source rocks** – muds, clays, marls and fine-grained carbonate rocks. The principal rocks capable of holding substantial volumes – **reservoir rocks** – to which they migrate are sandstones and carbonates. Minor quantities of hydrocarbons occur in Precambrian sedimentary rocks, though there are no commercial accumulations of undoubted Precambrian oil. Some large fields – notably those of Alberta, Canada – obtain oil from Palaeozoic reservoir rocks whose sources are most probably also Palaeozoic, but 80–85% of current production comes from Mesozoic or Tertiary rocks. The predominance of relatively young hydrocarbon accumulations is as likely to have resulted from the escape or destruction of older hydrocarbons as from changes in abundance of the parent materials. Hydrocarbons, notably methane, are among the volatiles emitted from some volcanoes, but such inorganic sources are generally discounted with respect to known oilfields.

The organic detritus deposited in hydrocarbon source rocks is thought to be derived from planktonic organisms (which multiply in the surface layers of seas and large lakes) where the supply of oxygen is inadequate. This detritus is buried in the sediment as a formless mixture of organic compounds known as **sapropel**. Most oil and gas reserves appear to have been derived from marine plankton. Oil formed in freshwater environments tends to be retained in the source rock, giving oil shales from which it can be recovered by distillation. Oil shales occur on a limited scale in Coal Measures in the Midland Valley of Scotland and on a huge scale in the Eocene Green River Formation of Wyoming, Utah and Colorado, which represents the deposit of a large lake formed within the developing Rocky Mountains. These deposits may yield one million barrels per day by the year 2000.

In marine environments, the anaerobic conditions needed for the preservation of sapropel result in part from high biological productivity in the surface waters and are favoured by basin forms which restrict the circulation of oxygenated water. The floor of the Black Sea today is partly covered by black sediments rich in organic matter (hence the term **euxinic**, from the Greek name for the Black Sea, as applied to anaerobic sedimentary environments). The source rocks of many oilfields accumulated behind reefs or other barriers or in narrow troughs defined by rifting or block faulting. The oil-bearing Mesozoic basins that flank the west coast of Africa, for example, appear to have originated when the Atlantic and Indian Oceans were represented only by narrow seas at the continental margins.

Under conditions of continued subsidence, organic debris buried beneath successive layers of sediment undergoes **diagenesis** along with the containing sediment. As temperatures and pressures rise, liquid hydrocarbons begin to segregate in the pore spaces. The reactions which form these compounds take place largely within the temperature range 40–150 °C, and the amount of oil generated is therefore controlled by the

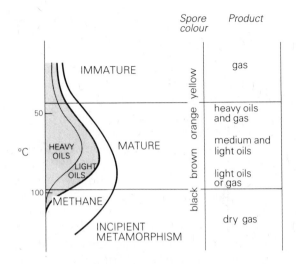

Figure 3.4 Maturation of hydrocarbons The temperatures given on the left represent approximately the maxima reached during burial.

geothermal gradient and the rate of subsidence. As an overburden of a few kilometres is required to reach the optimum temperature, thin successions generate little oil. On the other hand, rapid subsidence, or subsidence in regions of exceptionally high geothermal gradient, may carry the source rock through the critical zone into higher-temperature regions in which organic matter is degraded and the source rock is converted to carbonaceous shale or bituminous limestone. These reactions determine the depth to **economic basement**, defined as the level below which substantial accumulations are not to be expected. Economic quantities of oil seldom remain below 6–7 km depth, though methane may be generated at greater depths by devolatilisation of coal (Section 3.6.2). Some record of the thermal history of sedimentary rocks is provided by the diagenetic mineral assemblage and by modifications of relatively inert spores (Fig. 3.4).

3.5.3 Chemistry
Natural petroleum fluids are mixtures of many organic compounds which can be loosely divided into bitumen (too viscous to

flow readily), crude oil and natural gas. Viscosity, specific gravity and boiling point show a more or less continuous gradation from the relatively high values characterising the bitumen compounds to the low values of the more volatile oils (Table 3.2).

Hydrocarbons in the strict sense consist solely of carbon and hydrogen linked according to a few basic molecular patterns. Other organic compounds carry sulphur, nitrogen or oxygen in addition to carbon and hydrogen, and chlorine, vanadium, nickel, arsenic and other metals occur in trace amounts. Many of these minor components have a considerable nuisance value and are removed during processing.

Three principal hydrocarbon series are present – the paraffins, the naphthenes and the aromatics (Fig. 3.5). The paraffins, which are saturated with respect to hydrogen, have straight- or branched-chain molecules with the general formula C_nH_{2n+2}. Methane, CH_4, in which all four valencies of carbon are satisfied by hydrogen ions, may be thought of as the starting point of this series. Ethane, C_2H_6, is characterised by a two-carbon chain. Higher polymers with longer chains may form several **isomers** which differ in the number and position of side chains. Unsaturated analogues (the olefins) characterised by a lower H : C ratio and by double links between some adjacent carbons are formed during certain manufacturing processes; they may be converted to saturated hydrocarbons by addition of hydrogen.

Monocyclic naphthenes or cycloparaffins have the general formula C_nH_{2n} and may be likened to paraffin chains in which the two end carbons are linked into a closed loop, with the elimination of one hydrogen from each end. Cyclohexanes are based on a six-carbon ring, cyclopentanes on a five-carbon ring, both types being capable of embellishment by straight or branched side chains.

The aromatics (C_nH_{2n-6}) are based on the benzene ring C_6H_6, in which each carbon has two bonds with a neighbouring carbon (Fig. 3.5). These unsaturated hydrocarbons occur in nature and can be manufactured from naphthenes by a process of dehydrogenation.

Non-hydrocarbon compounds in crude oil and gas include those which carry sulphur and nitrogen. Sulphur forms from a fraction of 1% up to about 5%, varying from one oilfield to another. In light fractions it may be present as dissolved H_2S, but in the heavy fractions, where most of the sulphur resides, the element is combined with carbon–hydrogen chains or rings. Nitrogen occurs uncombined in natural gas but mainly in chain or ring compounds in oil. Vanadium and nickel occur in organic porphyrins structurally resembling chlorophyll; vanadium may reach over 1000 ppm, nickel over 100 ppm.

Table 3.2 Fuels derived from oil and gas (in order of boiling range).

natural gas (largely methane, CH_4)	gaseous at atmospheric temperatures, also occurs under pressure in solution in crude oil; usually piped directly from gasfield or liquefied for transport; domestic and industrial use, power stations
'bottled gas' (largely butane, C_4H_{10})	boiling range 10–0 °C, can be kept liquefied in steel containers at moderate pressures; mainly domestic use
light distillates	boiling range up to 160 °C, specific gravity ~0.74; gasoline, benzine, giving most motor fuel spirit and aviation fuel for piston-engined aircraft; high octane number attained by enrichment in aromatics and olefins
medium distillates	boiling range 160–350 °C, specific gravity ~0.80; kerosine, 'paraffin' (UK) and other grades giving diesel fuel, aviation fuel for jet aircraft, domestic and industrial boilers, power stations
heavy residual fractions	boiling above 350 °C, specific gravity ~0.83–0.95; heavy fuel oils used in ships, industrial boilers, power stations
waxes (large paraffin molecules)	candles, tapers

30

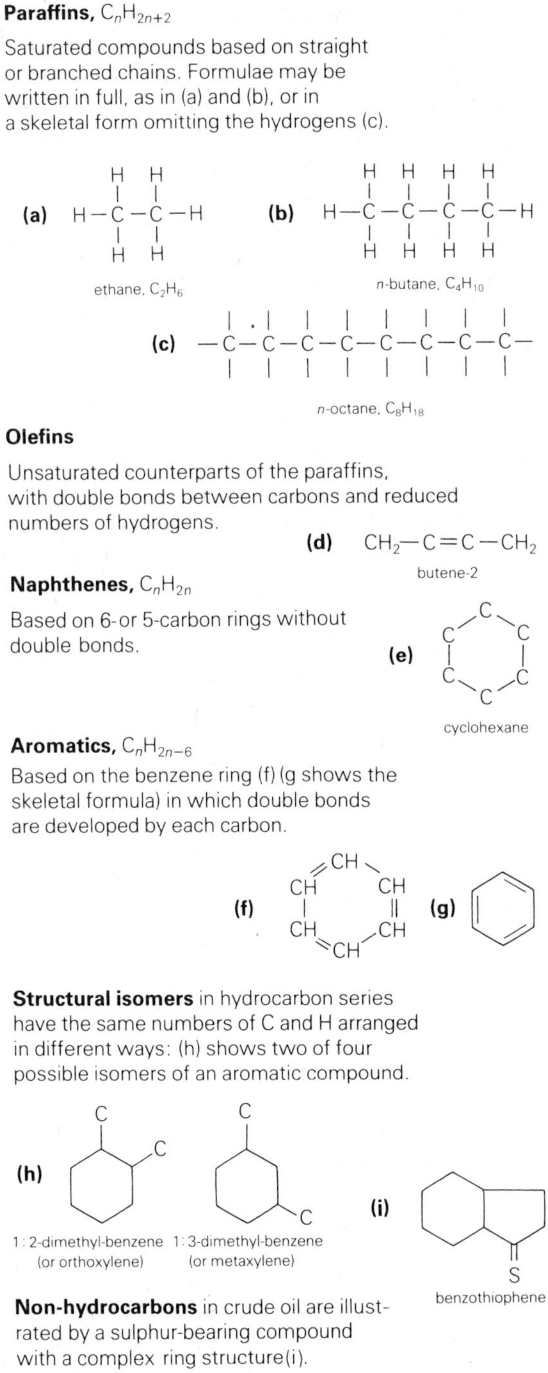

Paraffins, C_nH_{2n+2}

Saturated compounds based on straight or branched chains. Formulae may be written in full, as in (a) and (b), or in a skeletal form omitting the hydrogens (c).

(a)

ethane, C_2H_6

(b)

n-butane, C_4H_{10}

(c)

n-octane, C_8H_{18}

Olefins

Unsaturated counterparts of the paraffins, with double bonds between carbons and reduced numbers of hydrogens.

(d) $CH_2-C=C-CH_2$

butene-2

Naphthenes, C_nH_{2n}

Based on 6-or 5-carbon rings without double bonds.

(e)

cyclohexane

Aromatics, C_nH_{2n-6}

Based on the benzene ring (f) (g shows the skeletal formula) in which double bonds are developed by each carbon.

(f) (g)

Structural isomers in hydrocarbon series have the same numbers of C and H arranged in different ways: (h) shows two of four possible isomers of an aromatic compound.

(h)

1 : 2-dimethyl-benzene   1 : 3-dimethyl-benzene
(or orthoxylene)   (or metaxylene)

(i)

benzothiophene

Non-hydrocarbons in crude oil are illustrated by a sulphur-bearing compound with a complex ring structure (i).

Figure 3.5 The composition and structure of hydrocarbons

The associations and proportions of organic compounds in an oilfield determine the mode of processing and to some extent the uses to which the products may be put. In the refinery, the main types of fuel are separated initially by distillation, those 'cuts' with low boiling points being driven off first (Table 3.2). Natural gas consists predominantly of methane and paraffins up to butane (C_4H_{10}), and contains few impurities other than water vapour, nitrogen and minor amounts of CO_2, H_2S or He. Liquids of the lower boiling ranges from which gasoline is derived contain branched paraffins, cyclohexanes and aromatics, the last being valued because they yield products with high octane numbers. In the higher boiling ranges, straight-chain paraffins improve and aromatics detract from the quality of the fuel. Sulphur compounds are corrosive and are removed by various agencies. Heavy fuel oils used in industrial boilers and power stations retain a significant amount of sulphur which is oxidised to SO_2 during combustion and provides a major source of atmospheric polution (Section 7.6).

3.5.4 Migration and accumulation

Compaction and diagenesis lead to the expulsion of interstitial fluids from fine-grained sediments such as clays which undergo very large volume reductions. Since most petroleum source rocks are fine grained, these processes set in motion the first or **primary migration** of hydrocarbons into intercalated beds – usually sandstones – which retain relative high porosity and permeability during early diagenesis. Residual hydrocarbons in the parent rocks resemble those in the expelled fluid and may therefore 'fingerprint' the source of an oil pool.

Once transferred to permeable host rocks, oil and gas are capable of **secondary migrations** which are governed by factors similar to those controlling the movement of pore waters (Section 2.4). Permeable rocks of moderate to high porosity are capable of

31

transmitting and storing hydrocarbons; impermeable layers channel the flow and, when suitably arranged, provide barriers that obstruct further movement. Oil and gas, however, differ from water in three ways that have a bearing on the siting of oil and gas accumulations.

In the initial stages, oil fills only a part of the pore spaces, forming droplets immiscible with the associated pore water. Effective migration cannot take place until oil forms an interconnected system in the voids. The regional flow of water then generates a pressure gradient which sets the fluids in motion. Components of differing viscosity move with different speeds, the heavy hydrocarbons tending to remain nearer the source. The low specific gravity imparts a buoyancy to hydrocarbon fluids which leads them to accumulate above water in a permeable layer and to move to the highest part of a structural trap. Finally, and most important, migration of hydrocarbons is not linked to a cyclic circulation like that of water (Section 2.1). The destined routes are one-way, from source rock to surface, and enormous volumes of gas and oil escaping by way of surface seepages are lost by evaporation or oxidation. 'Lakes' of bitumen fed by seepages in Trinidad have been mined for asphalt (Section 5.2.7). The vast Cretaceous Athabasca oil sands of Saskatchewan lie immediately above an unconformity cut in oil-bearing Devonian limestones (Fig. 3.7). The oil which seeped into the McMurray Formation is heavy and viscous and cannot be pumped from the reservoir rock. Reserves estimated at well over 500 billion barrels might be exploited by 'mining' the sand wholesale using bucket excavators or other means.

Updip flow towards basin margins or positive areas within basins may transport hydrocarbons to rocks distinct in facies from those of their origin. In almost 60% of major oilfields, the reservoir rocks are sandstones; limestones, including reef limestones,

cavernous dolomites and jointed limestones provide most of the remainder. From what has been said above it will be clear that buoyant hydrocarbons may be trapped beneath an impermeable capping which seals off the fluid in the reservoir rock. The majority of **cap rocks** which serve this purpose are argillites, marls or evaporites (mainly gypsum).

Oil and gas traps are of many kinds (Figs 3.6, 7 & 8). **Stratigraphical** (or **lithological**) **traps** are essentially primary and they usually result from lateral variations in sedimentary facies. Near basin margins and structural 'highs' lenticular sandstones or reef-limestones may be capped by shales or evaporites. The sandstones of such environ-

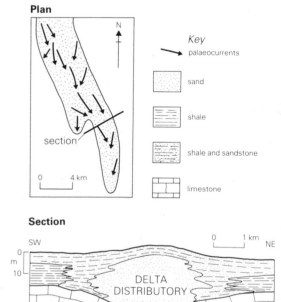

Figure 3.6 Oil traps Lithological factors in their development, illustrated by deltaic deposits in the Mississippian of Indiana. The permeable sand body shown in plan and section represents the fill of an ancient distributary. It has a maximum thickness of ~40 m and is bordered by and overlain by interbedded shales, silts and very fine sandstones. The bulge over the sand body is due to compaction (based on Hrabar, S. V. and P. E. Potter 1969. *Bull. AAPG* **53**, 2150).

ments may appear as basal beds in a transgressive series, as offshore banks, as bars separating open sea and lagoonal waters or in the distributory channels of deltas. Many bodies, known by the expressive term **shoestring sands**, are linear in plan, paralleling a shoreline or delta distributory channels (Fig. 3.6).

Reefs are calcareous mounds or ridges built by colonial corals, bryozoa and other rock-forming organisms. They develop in warm, clean, shallow marine waters, often fringing the shoreline and backed by lagoons in which fine detritus and reef debris accumulate. Reef limestones appear as lenticular or elongate bodies interfingering with the finer sediments on either side and draped by overlying strata in such a way as to produce

stratigraphic traps. The effective porosity and permeability of reef rocks and associated sediments depend as a rule on diagenetic processes of replacement, solution and deposition rather than on the initial (usually very high) porosities. Initial voids may be infilled but subsequent dolomitisation may cause a reduction of volume and development of further voids and pores. Reef complexes make suitable hosts not only for oil and gas but also for ores of lead and zinc deposited from migrating brines (Section 4.4.3). Both types of accumulation are represented in Devonian reefs developed parallel to northeasterly structural 'highs' in Alberta (Fig. 3.7).

Structural traps are produced by post-depositional tilting, folding and faulting

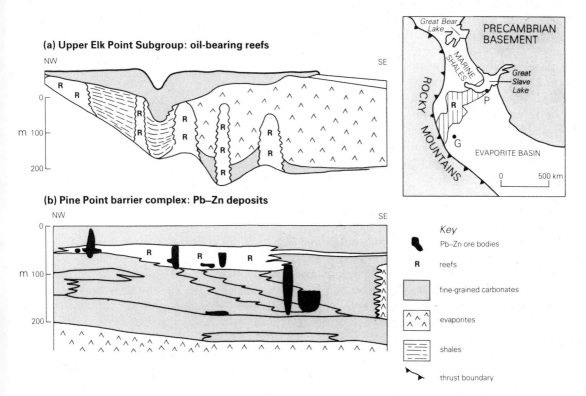

Figure 3.7 Reefs (a) As reservoir rocks for oil. (b) As hosts for lead–zinc deposits. The sketch map shows the position of a Middle Devonian reef barrier separating open sea to the north from an enclosed evaporitic basin to the south, in northwestern Canada. The Goose River oilfield (G) is located in a subsidiary reef tract (section illustrating facies variations based on Barss, D. L. *et al.* 1970. *AAPG Memoir* no. 14). The Pine Point complex is referred to in Section 4.4.3 (details based on Kyle, J. R. 1981. *Handbook of stratabound and stratified ore deposits III.* Amsterdam: Elsevier).

(Fig. 3.8). **Anticlinal folds** and **domes** in which oil or gas is trapped under an impervious cap rock can often be detected by surface mapping or by geophysical (seismic, gravity) surveys and have been the first target of many exploration programmes. The majority of productive anticlinal structures, such as those of the Middle East (Section 3.5.6, Fig. 3.11), are comparatively simple; intense folding in sandstones or limestones tends to lead to rupture and hence to loss of oil via fractures. Knowledge of the three-dimensional geometry of the folded reservoir bed is essential if wells are to be correctly sited.

Evaporites in sedimentary successions (Section 5.5) are of importance to the oil geologist in at least two respects. They are largely impervious and therefore provide excellent cap rocks, and their low density and high ductility give them a tendency to rise relative to other sedimentary rocks, thereby creating domes and ridges which can act as traps. The role of evaporites as seals is enhanced by the fact that **sabkha belts**, in which salts are precipitated from interstitial brines permeating the accumulating sediments, may be located in the vicinity of basin margins towards which the hydrocarbons tend to move. The effects of **salt tectonics** or **halokinesis** in tectonically stable regions influences the distribution of oil, as in the southern states of the USA and the southern North Sea. Salt-controlled structures range from simple domes and arches, in which the overlying strata are draped above the buoyant salt mass, to complex **diapirs** around which the overlying strata are disrupted and pierced by a plug, ridge or mushroom-shaped salt intrusion. Oil or gas may occupy the domed roof region or the uptilted parts of ruptured beds (Fig. 3.8).

Oil traps may result from **faulting** where permeable and impermeable horizons are juxtaposed, as in the vicinity of the Viking and Central Graben of the North Sea (Fig.

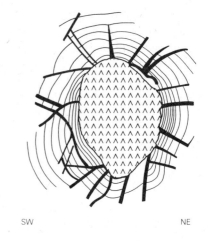

SW   NE

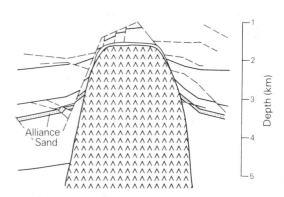

Figure 3.8 Salt dome trap Uplifting and rupturing of the Alliance Sand of the Guedan Dome, Louisiana, is illustrated in plan (above), showing form lines on the top of the sand and radial faulting, and in section (below) (based on Murray, G. E. 1968. *AAPG Memoir* no. 8).

3.13). Faulting at depth may be expressed by monoclines or 'drape' structures at higher levels. A fundamental but indirect connection with faulting arises from the fact that many oilfield successions accumulated in fault-bounded basins. The North Sea oilfields are clustered along a branching rift system (Fig. 3.13) and the oil basins of Angola, Gabon and East Africa correspond to Mesozoic horst and graben zones developed at fracture coastlines prior to the disruption of Gondwanaland.

3.5.5 Prospecting and development

The search for oil begins with the recognition of a potentially interesting region and, with good fortune, proceeds to the identification of productive targets for detailed assessment. The general characters to be sought will be clear from what has been said in earlier sections: thickish Phanerozoic marine successions of shelf seas, especially those deposited in basins with restricted bottom circulation, which incorporate both fine-grained potential source rocks and permeable sandstones or carbonate rocks. In the western hemisphere, at least, most assemblages of this type in land areas have already been explored and primary investigations have shifted mainly to offshore regions.

When the prospects of success seem good after preliminary photogeological, geophysical and geological reconnaissance, test boreholes may be drilled on a favourable-looking structure. If oil is proved, the investigation moves into a new phase of drilling and geophysical surveying aimed at locating individual oil-bearing structures and building up a regional picture of variations in lithology, diagenetic state and structure. The techniques used to establish the three-dimensional forms of folds, salt domes, fault traps and stratigraphical traps include the standard methods of structural analysis based on the projection of observations made at the surface; gravity surveys, especially useful where low-density evaporites are present; and seismic reflection and refraction surveys designed to identify marker horizons at depth. These exploration methods are outlined in Chapter 8, but two methods which have been brought to fine arts in the oil industry deserve further mention here. Seismic exploration on a regional scale leads to the recognition of what has been called a **seismic stratigraphy** in which 'packets' of genetically related strata are bounded above and below by structural discordances which are easily picked out on seismic sections (Fig. 3.9). The distinctive seismic properties (velocity, presence or absence of internal reflectors) of each packet or **depositional sequence** facilitate correlation. The discordant boundaries generally mark phases of transgression, regression or erosion which can be dated palaeontologically from borehole samples.

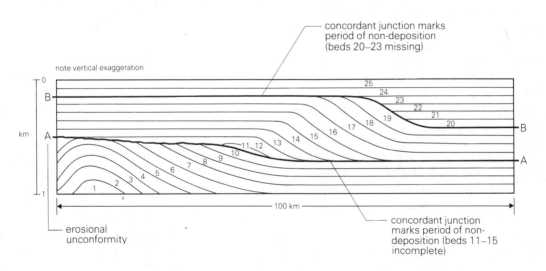

Figure 3.9 Depositional sequences Three sequences defined by seismic profiling. The bounding surfaces A and B are defined by minor unconformities recognisable on seismic sections (based on Mitchum, R. M., P. R. Vail and S. Thompson 1977. *AAPG Memoir* no. 26).

Techniques based on drilling yield a variety of information (almost 1800 wells were drilled in 16 years of North Sea exploration and development). Although drill cores that are recovered intact give additional information, most operations yield only chips carried up in the lubricant circulated in the borehole. The tedious but essential process of mud logging provides basic information about the lithology and diagenetic state of the rocks encountered. The physical properties of the rocks traversed may be recorded automatically on the drilling time log or by instruments lowered down the hole before its sides are cased in. The most important of these ancillary records are Schlumberger logs of resistivity and other electrical properties which vary with the porosity of the rocks and the composition of their pore fluids; interstitial brines are good conductors, compared with fresh water or hydrocarbons. Logs recording natural radioactivity show anomalies corresponding to high levels of K, Rb, or U which generally indicate argillaceous rocks. Schlumberger and radiometric logs provide data that assist well-to-well correlation, the same sequence often showing remarkably similar features in widely separated localities (Fig. 3.10, Table 3.3).

Intensive investigation of individual sites is needed to estimate the probable size and structure of an oil pool. Such an exercise requires data on (i) the three-dimensional form and distribution of the reservoir rock, (ii) variations in porosity, and (iii) the depth and inclination of the oil/water interface. Only a fraction of the oil in pore spaces can ultimately be extracted and the recovery factor (see below), which determines the yield, varies from oilfield to oilfield. Published figures for 'proven' reserves (e.g. Table 1.2) are based on appraisals of the total volume and probable recovery factor. Where adequate reserves are proved, production can begin. Initial fluid pressures may be sufficient to force hydrocarbons to the sur-

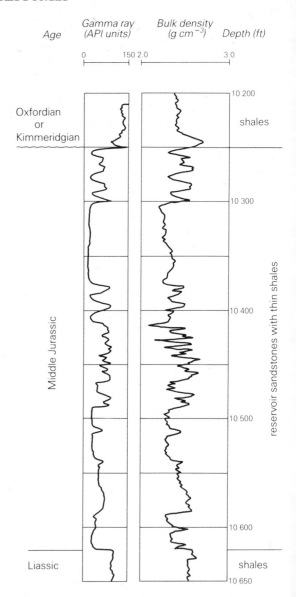

Figure 3.10 Information from borehole logging An example from the Ninian field, northern North Sea (based on Albright, W. A. 1980. *AAPG Memoir* no. 30).

face; at later stages, fluids must be pumped up for transmission to refineries. In offshore fields, the well head is linked to large production platforms capable of standing on the sea bed in waters up to about 150 m.

In its natural condition, oil trapped in a reservoir is under pressure. When a well

Table 3.3 Information obtained during drilling.

continuous cores	lithology, porosity, permeability, thickness of units, stratigraphical position (fossils), angle of dip, traces of hydrocarbons
rock chips	lithology, porosity, permeability, stratigraphical position (microfossils), traces of hydrocarbons
rate of penetration	hardness of rock
drilling mud	hydrocarbons in mud suggest oil or gas in pore fluids: viscosity varies with temperature: mud lost in hole indicates fissured or very permeable rocks
temperature in hole	heat flow/geothermal gradient
downhole logging Schlumberger logs of electrical resistivity, conductivity, etc.	porosity, composition of pore fluids (hydrocarbons poor conductors, brines good conductors)
radiometric log	high natural radioactivity suggests shales; secondary radioactivity related to porosity
sonic log (speed of sound in rock)	porosity

penetrates oil or gas, precautions are taken to prevent the explosive release of volatile hydrocarbons coming out of solution, or of methane and other gases at the top of the reservoir. As oil (or gas) is drawn off during production, further oil moves into the base of the well at a rate which depends on the viscosity of the oil and the permeability of the host rock. These factors determine the rate of production which is higher for light oils and for reservoir rocks in which flow takes place mainly through fractures rather than pore spaces (see Asmari Limestone, Section 3.5.6). The upward movement of oil during the primary production stage is facilitated by its buoyancy, by expansion of the water beneath the oil pool which follows reduction of pressure, and by the effects of gas exsolving under reduced pressure. The **recovery factor** is expressed as the percentage of hydrocarbon that can be extracted. Only some 20–40% of the oil *in situ* is usually recovered by primary production

methods. Systems of secondary recovery, which have significantly increased the yield in some oilfields, involve the displacement of a further fraction by water pumped down into the reservoir; this process of water flooding is analogous to the artificial recharge of aquifers depleted by excessive pumping (Section 2.6.1). Recovery of natural gas is higher than that of oil, rising to 90% in some gasfields.

3.5.6 The Middle East

Bitumen has been used as a building material from prehistoric times and the first recorded oil well was dug about 500 BC in the Persia of Darius the Great. At the present day, Middle Eastern oilfields not only provide over a third of world production (Table 3.4) but also carry well over half of the proven reserves (Table 1.2). They lie in a crescentic zone which extends through Iraq, Iran, Kuwait and Saudi Arabia to the Persian Gulf and the Gulf of Oman, where offshore fields are worked from platforms (Fig. 3.11). From comparatively small beginnings between the two world wars, production has risen rapidly, and by 1973 annual production approached 8000 million barrels (336 000 million gallons).

The oilfields lie near the northern margin of the former supercontinent of Gondwanaland in an area flanked to the north-east by the Alpine–Himalayan mobile belt developed in Cretaceous to Tertiary times (Fig. 3.11). In the foreland region and marginal part of the mobile belt, shallow-water deposition on a relatively stable basement continued with little interruption from

Table 3.4 Production of oil, 1973.

	million barrels per day
world total (estimated)	56.7
USA	8.9
Middle East	20.7
Saudi Arabia	8.4
USSR (estimated)	9.2

One barrel = 35 Imperial gallons or 42 US gallons.

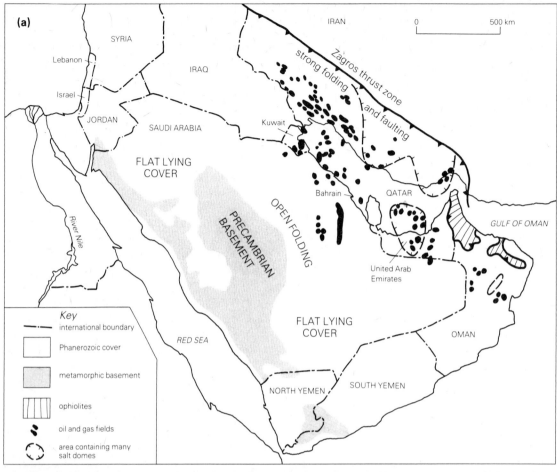

(a)

SYRIA

Lebanon

Israel

JORDAN

IRAQ

IRAN

0   500 km

Zagros thrust zone

strong folding and faulting

Kuwait

SAUDI ARABIA

FLAT LYING
COVER

PRECAMBRIAN
BASEMENT

OPEN FOLDING

Bahrain

QATAR

GULF OF OMAN

United Arab
Emirates

River Nile

FLAT LYING
COVER

OMAN

RED SEA

NORTH YEMEN   SOUTH YEMEN

Key
— ⋅ — ⋅ international boundary
☐ Phanerozoic cover
▨ metamorphic basement
▥ ophiolites
🝔 oil and gas fields
⬭ area containing many
salt domes

(b)

Figure 3.11 Oilfields of the Middle East (a) Their location in relation to the geology of the area. (b) A limestone anticline in Iran, illustrating the form of anticlinal traps in the Middle Eastern oilfields (Aerofilms).

Carboniferous to mid-Tertiary times. The supply of detritus was limited and the principal deposits are limestones, argillaceous rocks and evaporites. Towards the interior of the Afro-Arabian craton, the sequence shows little disturbance apart from that associated with salt tectonics. Towards the north-east, large and relatively simple folds provide immense anticlinal oil traps (Fig. 3.11b). The northeastern limit of the main producing areas is the Zagros Range, a linear zone of intense orogenic disturbance which may mark a former plate suture or transform dislocation.

In terms of geological setting, the Middle Eastern oilfield zone has much in common with other continental margins that remained stable and submerged through long periods of time. Its exceptionally high content of hydrocarbons appears to reflect an unusual combination of favourable circumstances:

(a) The detrital marine horizons included one or more units rich in sapropel; the principal source rocks of oil were probably Mesozoic in age.

(b) The source rocks were interleaved with permeable strata capable of storing hydrocarbons; the principal Mesozoic reservoir rocks are Upper Jurassic and Lower Cretaceous carbonates (Thamama Limestone); Upper Cretaceous limestones and late Cretaceous sandstones carry lesser quantities and a little gas (probably derived from underlying coals) is contained in Permian strata; the major Tertiary reservoir rock – the main producer in Iran – is the Oligocene–Miocene Asmari Limestone. As the primary migration took place early, hydrocarbons in the pores inhibited cementation and high porosities were retained, the Wasia Sandstone formation (Albian–Cenomanian) in Saudi Arabia, for example, has a porosity of about 25%.

(c) The long history of almost uninterrupted sedimentation favoured the retention of hydrocarbons since there were few structural disturbances or phases of erosion which might promote escape. Shales, evaporitic marls or evaporites overlie several reservoir rocks, the most important of these being the Upper Jurassic clays and evaporites which overlie productive horizons in Saudi Arabia and parts of the Persian Gulf, Eocene evaporitic sequences overlying Upper Certaceous reservoirs and mid-Tertiary (Fars) marls, evaporites and shales which overlie the Asmari Limestone in Iran.

(d) Broad folding of the thick, competent limestones in middle and late Tertiary times gave large upright or inclined anticlines continuous along the axial direction for many tens of kilometres. Fracturing near fold crests increased the permeability of the reservoir rocks and is responsible for the exceptionally high productivity of certain wells. Wells in fractured Asmari Limestone in Iran may yield as much as 25 000 barrels per day (cf. transmission of water, Section 2.4).

(e) Despite the tectonic disturbances resulting from collision of the Afro-Arabian and Eurasian blocks, the oilfield region, unlike the Zagros and Alborz Mountains on its northeastern flank, was neither severely disrupted nor deeply eroded; large parts of the Asmari Limestone and most of the Mesozoic limestones remain below the level of erosion and are still sealed by the overlying cap rocks.

3.5.7 Southern USA

The United States was for many years the world's largest single producer and consumer of hydrocarbons, producing over 18% and consuming almost 30% of the world totals in 1973 (Table 3.4). Drilling began in the mid-19th century in the southern Appalachians where oil seepages were well known, but the more important oilfields of the southern states and California, which between them account for over 65% of the present annual production, were not found until late in the century. With the exception of the Alaskan and offshore fields, the principal oilfields of the USA have by now been thoroughly explored and large new finds are becoming rare. The known reserves are still large and are being increased by secondary recovery schemes; and large, so far little-used, reserves of oil shale are available (Section 3.5.2). As the discrepancy between production and consumption shows (Fig. 1.4), however, the United States is likely to remain an importer of hydrocarbons for the forseeable future.

The oilfields of Kansas, Oklahoma, Texas and Louisiana are, like those of the Middle East, located close to an old continental margin in a region subject to net subsidence over a long time period (Fig. 3.12). There, however, the resemblance ends. The American fields are located in predominantly detrital successions. Those of West Texas,

Kansas and Oklahoma extract Palaeozoic oil from Palaeozoic sandstones, Mississippian limestones and Permian reef complexes. The cap rocks include shales and evaporites, the latter sealing off Permian reef-limestones in West Texas. Oil is contained partly in wedge-shaped sandstones and linear reef complexes and partly in broad domes and faulted arches formed in response to block movements of the basement.

The oilfields of the Gulf Coast of Texas, Louisiana and Mississippi are situated in a great wedge of marine and non-marine Mesozoic to Tertiary sediments built out towards the Gulf of Mexico by forerunners of the Mississippi. The reservoir rocks are almost entirely clastic sediments among which shoestring delta distributaries provide stratigraphical traps sealed by fine detrital sediments. Permian to Triassic evaporites underlying the productive sequence over a vast area and extending beneath the floor of the Gulf provide numerous salt domes and diapirs on which many small oilfields, both onshore and offshore, are sited (Fig. 3.8).

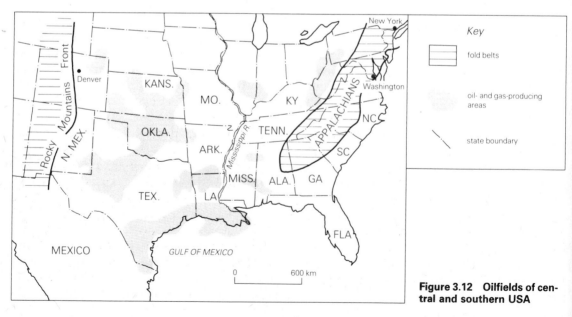

Figure 3.12 Oilfields of central and southern USA

3.5.8 The North Sea

Exploration and development in the North Sea – involving costly and technically difficult operations in notoriously stormy waters – were delayed until the second half of the 20th century. Production of gas began in the 1960s and of oil in the mid-1970s. Oil output from the British sector is expected to reach two to three million barrels per day in the 1980s, making the United Kingdom self-sufficient for a time, but will decline towards the end of the century. The full extent of the reserves has yet to be established.

The oilfields lie in the heavily fractured northwestern part of the European craton which has suffered no orogenic disturbance since the end of the Lower Palaeozoic (Fig. 3.13). Intermittent deposition since late Palaeozoic times has given a sequence of predominantly marine Mesozoic and Tertiary sediments amounting in places to well over 4 km in thickness. Subsidence was largely fault controlled, the bulk of the sediment being contained in a central branching system of graben or half graben formed during the period of crustal extension which preceded the opening of the Atlantic Ocean. The basin is partially divided into northern and southern portions by structural 'highs' which extend irregularly from north-east England to Denmark (Fig. 3.13).

Natural gas, without related oil, is found in the western part of the southern North Sea where well sorted Lower Permian (Rotliegendes) aeolian sandstones representing a dune belt on the flanks of a wide shallow lake provide the reservoir rocks, and Zechstein evaporites the cap rocks. The source rocks are thought to be Carboniferous coals unconformably underlying the Permian.

Oil, with or without associated gas, occurs more widely in both southern and northern basins. It is confined to Mesozoic and Tertiary rocks and is concentrated mainly near the central rift system. Shales of late Jurassic (Kimmeridgian) age are regarded as the principal source rocks, and sandstones distributed through the sequence (sealed by interleaved shales) the main reservoir rocks (cf. Fig. 3.10). Stratigraphical traps are provided by lenticular sandstones, or by beds truncated by low-angle unconformities, and structural traps mainly by faults or (near the southern shore) by ridges and domes of Permian salt.

From Triassic to mid-Cretaceous times, variations in thickness and lithology were related to the development of the central graben and the associated block faults (Fig. 3.13); this phase was accompanied by Jurassic volcanic activity at the intersection of northwesterly and northerly graben. It ended with an episode of tectonic disturbance during which the contents of the linear fault basins were uparched and partially eroded, a process often referred to as 'inversion' of the basins. The succeeding Upper Cretaceous and Tertiary strata were laid down as a blanket, thickening over parts of the former rift to as much as 3 km.

3.6 Fossil fuels: the coal series

3.6.1 General characteristics

Solid carbonaceous rocks of the coal series which fuelled industrial development of Europe and North America in the 19th and early 20th centuries still provide for over 20% of energy consumption in the USA, around 40% in western Europe, and 5–10% of the feedstock for the western chemical industry. World production ran at over 2400 million tonnes in 1981. Known reserves in Europe, North America and the USSR are not likely to be exhausted for two to three centuries and production is expected to outpace that of hydrocarbons in the 21st century. The restrictions on production are at present social and economic rather than geological. Opencast mining, which could be extended, especially in the eastern USA, leaves a scarred countryside stripped of coal and soil and

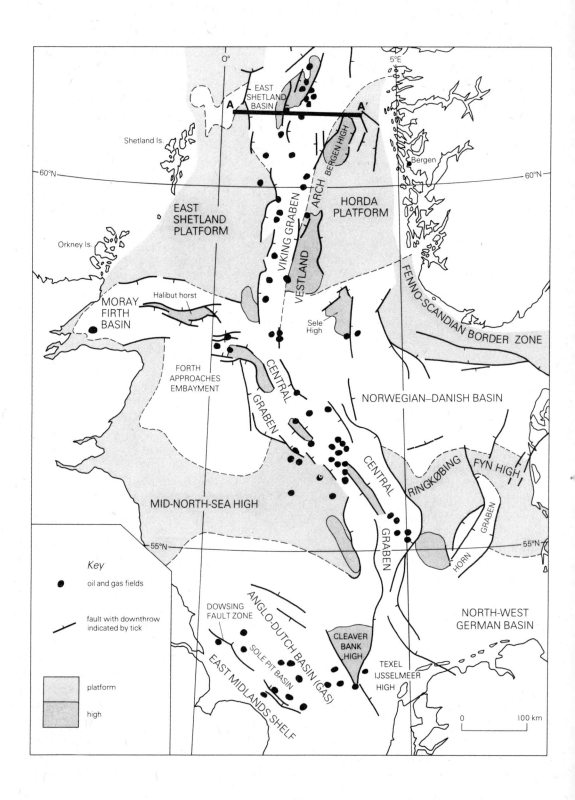

5°E

0°

EAST
SHETLAND
BASIN

A ————————————————————— A'

Shetland Is.

Bergen

60°N   60°N

EAST
SHETLAND
PLATFORM

VIKING GRABEN

BERGEN HIGH

VESTLAND ARCH

HORDA
PLATFORM

FENNO-SCANDIAN BORDER ZONE

Orkney Is.

MORAY
FIRTH
BASIN

Halibut horst

Sele
High

FORTH
APPROACHES
EMBAYMENT

CENTRAL GRABEN

NORWEGIAN–DANISH BASIN

55°N   55°N

MID-NORTH-SEA HIGH

CENTRAL GRABEN

RINGKØBING FYN HIGH

HORN GRABEN

NORTH-WEST
GERMAN BASIN

DOWSING
FAULT ZONE

ANGLO-DUTCH BASIN (GAS)

SOLE PIT BASIN

CLEAVER
BANK
HIGH

TEXEL
IJSSELMEER
HIGH

EAST MIDLANDS SHELF

Key

● oil and gas fields

fault with downthrow
indicated by tick

platform

high

0   100 km

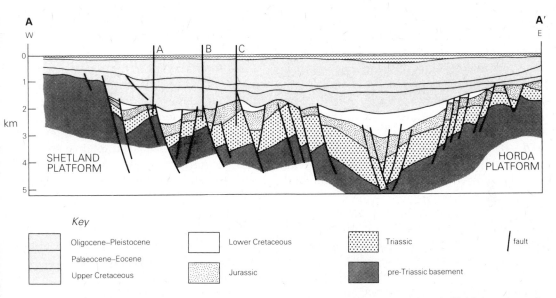

Figure 3.13 The structure of the North Sea oilfields The map illustrates the concentration of oilfields in and near the graben. The section illustrates the structure of the Viking graben, northern North Sea (section A–A′); note the contrast between the heavily fractured Triassic–Lower Cretaceous and less disturbed Upper Cretaceous–Pleistocene. Wells A, B and C recover oil from fault traps (section based on Kirk, R. H. 1980. *AAPG Memoir* no. 30).

requiring extensive rehabilitation; and underground mining, the principal method used in Britain, calls for the employment of labour in unpleasant and often dangerous conditions. The traditional methods that involve bodily removal of coal to the surface have already been transformed by mechanisation and in some instances are being superseded by the conversion of coal to a slurry that can be pumped to the surface. Methods involving the underground conversion of coal to gas or the underground generation of electricity may be extensively used in the future. Production of coal in Britain was running at about 125 million tonnes per annum at the end of the 1970s.

The **coal series** comprises all the solid carbonaceous rocks formed as a result of burial and diagenesis of large volumes of plant debris. Organic matter is supplied mainly by land vegetation and represents the litter accumulated on forest floors and in associated swamps and watercourses. The

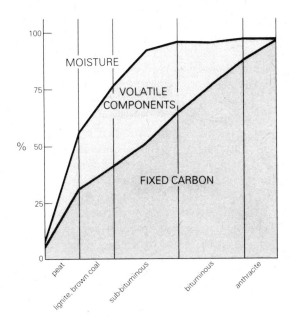

Figure 3.14 The coal series

partially rotted woody and soft tissues (which differ from sapropel in being rich in carbohydrates) build up *in situ* as peaty layers with low densities and high porosities. Diagenetic changes lead to compaction, with expulsion of water and volatiles, an increase in carbon content and density and a corresponding reduction in volume and porosity. By these means original debris may be progressively converted to seams of peat, lignite, brown coal, bituminous coal and anthracite (Fig. 3.14).

The distribution of the coal series in time and space reflects evolutionary, climatic and geological controls (Fig. 3.15). As the parent materials are derived from advanced land plants, coals are virtually restricted to Upper Palaeozoic and post-Palaeozoic successions, deposited after the colonisation of the land by plants. Occasional carbonaceous layers in older rocks, such as the anthracite-like shungite in early Proterozoic successions of the Baltic Shield are probably of algal origin (cf. torbanite, Section 3.6.2). By a coincidence that has not been fully explained, a large proportion of the world's coal was formed over a timespan of only some 75 million years towards the end of the Upper Palaeozoic era when coal forests spread widely in Europe, North America and the supercontinent of Gondwanaland. Mesozoic coals are important in Siberia, China and along the east flank of the Canadian Rockies, and Tertiary coals in East and West Germany and the USSR. Peat formed in the aftermath of the Pleistocene ice age is a traditional fuel, especially in Scotland and Ireland. As might be expected, lower-rank members of the coal series are common in the younger systems, whereas Upper Palaeozoic coals are generally of high rank.

Most coal forests appear to have been situated in warm-temperate to subtropical regions of moderate to high rainfall where climatic conditions favoured rapid growth and where a diversity of species could grow. The Permo-Carboniferous coals of Gondwa-

naland, however, overlie or are interstratified with glacial sediments and may have been formed in cool-temperate to sub-arctic climates analogous to those under which the postglacial peats of northern Europe accumulated. The Gondwanaland coal flora appears to have been dominated by a single genus, the simple-leaved *Glossopteris*.

The low-lying coastal plains and deltas most favourable to the growth of extensive coal forests are sensitive to even minor changes in relative sea level and the prolonged accumulation of organic material therefore required periods of tectonic stability. The characteristic rhythmic variations in

Period	Time (million years)	
Quaternary and Neogene		peat extensive in sub-arctic regions Germany, Ukraine, Baikal region (lignite, brown coal)
Palaeogene		C. Europe, Ukraine, USA (lignite, brown coal)
Cretaceous	100	Siberia, N. Germany, Canadian Rockies
Jurassic		Siberia, China; locally C. Europe (brown coal)
Triassic	200	Yunnan (China); locally eastern USA
Permian		Gondwanaland coals in S. Africa, Brazil, Australia, India; Siberia, China (mainly bituminous)
Carboniferous	300	Upper main coal basins Silesia, Westphalia, Belgium, France, Britain, USA (bituminous), Siberia
		Lower Moscow Basin (lignites)
Devonian		land floras diversify
	400	
Lower Palaeozoic		no advanced land flora
	500	

Figure 3.15 Major coal-bearing formations Their stratigraphical and geographical distribution.

Table 3.5   Coal measure cyclothems.

Extended cyclothem in East Pennine area (England)

coal	simplest cycles
seat earth=palaeosol	consist only of these
shale, non-marine	units
siltstone and sandstone	
shale, non-marine	
(base) shale, marine (not present in all cycles)	

A representative cyclothem in Ohio (USA)

coal
clay, non-marine
shale and sandstone, largely marine
(base) limestone, marine

In American literature, coals are usually taken as the basal units of the cyclothems: the sequence cited above is transposed into the same form as the British cyclothem.

coal-measure successions, and the frequent interleaving of non-marine and marine sediments, are evidence of the delicate balance maintained in these **paralic zones** (Table 3.5). Smaller **intermontane basins** gave rise to less extensive Carboniferous coalfields as at St Etienne in the central massif of France.

3.6.2   Raw materials and end products

The parent materials of the Palaeozoic coals came from spore-bearing plants – ferns, lycopods, horsetails and others – and seed-bearing plants of early groups now reduced or extinct. Mesozoic and Tertiary coals (Fig. 3.16) were derived mainly from gymnosperms and angiosperms allied to modern groups of seed-bearing plants. All these groups contributed similar organic compounds to the forest litter. Lignin, the main component of wood, and cellulose, the substance of plant cell walls, are carbohydrates characterised by complex molecules containing carbon, oxygen and hydrogen. Leaf cuticles, pollen and spores are formed by the waxy hydrocarbon cutin, made essentially of carbon and hydrogen. The organic matter is

Figure 3.16   Tertiary lignites   Note the small faults which interrupt the continuity of these seams near Newton Abbot, south-west England (Crown Copyright reserved).

attacked by fungi and by aerobic and anaerobic bacteria, with the liberation of carbon dioxide, water, hydrogen and methane. Under reducing conditions, stagnant ground waters are enriched in humic acid and other organic solutes which halt bacterial action before decomposition is complete. The partly rotted plant matter is then impregnated with organic acids, alcohols, fatty compounds and hydrocarbons, often mixed with sulphides generated by bacterial action. Inorganic detritus introduced by wind or streams contaminates the mixture and provides incombustible **ash** in the final product. Elements such as uranium and vanadium that are easily fixed under reducing conditions may be precipitated in significant amounts (Section 3.4.2).

Compaction and diagenesis lead to dewatering, the expulsion of light hydrocarbons and the formation of solid organic complexes from the impregnating solutions (Table 3.6, Fig. 3.14). Total carbon contents rise from well below 50% to over 90% and the carbon : oxygen ratio increases progressively. The light hydrocarbons (principally methane CH_4) expelled during diagenesis migrate from the source rock and may be trapped elsewhere as reservoirs of 'dry' natural gas (cf. Section 3.5.8). Coals of low to medium rank can be made to yield further hydrocarbons when burned under reducing conditions, a process used since the early 19th century to manufacture coal gas for lighting and heating; the spongy residue of combustion, coke, provides a smokeless fuel formerly much used in power stations. Sulphur and other impurities are released on burning, with consequent pollution of the atmosphere (Section 7.6).

The great majority of commercially used coals are humic coals deposited *in situ* and derived from mixed plant debris (Table 3.6). At high ranks, the humic coals are lustrous and usually well laminated, and they break into joint-bounded, often rectangular, pieces. The carbohydrate complexes and other materials of which they are composed rarely preserve traces of cell structure. Lamination due to the alternation of the components listed in Table 3.6 is observable in hand specimens and is emphasised under the microscope by differences of reflectance (Appendix 1). Sapropelic coals are relatively rich in hydrocarbons and may be derived both from the spores or pollen of higher plants and from algal remains. Cannel coal, resulting mainly from deposition in stagnant pools, is a minor component of most coal measures. Boghead coal is partly, and torbanite mainly, derived from algal matter. Sapropelic coals have a greasy lustre and break with a conchoidal fracture. They ignite easily at low temperatures and yield gas and oil on distillation.

Table 3.6 Components of high-rank coals.

Organic

In humic coals:
(a)   clarain – finely laminated mixture of three components:
    vitrinite – black, shiny, carbohydrate-rich plant tissues impregnated by hardened gels derived from organic decomposition produces;
    micrinite – black, dull, derived from fine debris including spores, cuticles and other hydrocarbon materials;
    fusinite – black, powdery essentially fine charcoal, occurring as films which provide parting planes.
(b)   durain – hard, massive, lustreless, composed mainly of micrinite and fusinite.

In sapropelic coals:
(a)   cannel – comparable with durain, derived from fine debris including hydrocarbons;
(b)   boghead – with significant proportion of algal matter;
(c)   torbanite – derived from algal matter.

Inorganic ash
detrital sand, silt, clay, carbonates, sulphides, adsorbed U, V, etc.

3.6.3   Coal Measures of Europe and North America

The late Palaeozoic Coal Measures, which extend through Europe and North America at the northern margin of the Hercynian – Appalachian orogenic belt, provide one of the world's largest assemblages of coals and illustrate the geological factors governing their development. The principal phase of deposition (Upper Carboniferous in European, Pennsylvanian in American terminology) came during the late stages of orogenic activity when the rising Hercynian mountain belt shed detritus onto the paralic plains flanking its northern foreland (Fig. 3.17). The bulk of the succession consists of non-marine shales and sandstones which are interleaved on the one hand with marine bands deposited during short-lived transgressions and on the other with coal seams and ancient soils (seat earths). Essentially the same sequence of rock types is repeated many times in characteristic rhythmic units or **cyclothems** averaging about 20 m thickness (Table 3.5). Each cyclothem ends with a coal formed during a phase of stability and is succeeded at the base of the next unit by detrital, sometimes marine, sediments recording a significant change of base level and/or sediment supply that points to a tectonic control. Individual cyclothems persist

laterally over wide areas and can be identified in exploratory boreholes by reference to their lithological and faunal peculiarities. The same four principal marine bands, for example, appear in most British coalfields and some can be identified as far east as Germany. Regional variations in space and time reflect the palaeographical evolution of the basin. In the United States, for example, marine bands are more numerous near the border of the Appalachians than westwards towards the interior of the continent. In Britain, marine bands are more important in the lower part of the succession deposited when the sea occupied a large part of the Hercynian mobile belt than in the upper divisions laid down when orogenic uplift had taken place.

Although many coal seams or groups of seams have a wide lateral extent, both primary and secondary complexities may affect their suitability for mining. A normally workable coal may split into two or more

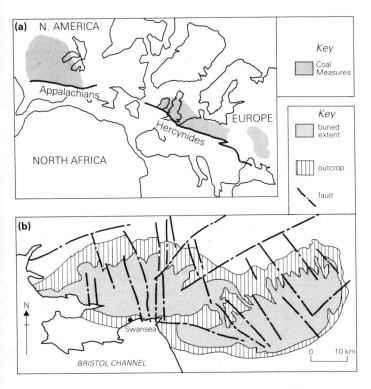

Figure 3.17 Upper Carboniferous (Pennsylvanian) Coal Measures (a) The extent of coal forests in the supercontinent of Laurasia in relation to the Appalachian and Hercynian mountain belts. (b) The coal basin of South Wales (UK) showing the outcrop and subsurface extent of the productive Lower and Middle Coal Measures. The coalfield lies immediately north of the Hercynian orogenic front and is strongly disturbed. Its structure is that of an asymmetrical syncline, with a relatively steep southern limb, disrupted by two sets of transcurrent faults.

thin seams where unusually rapid subsidence led to the drowning of forests and inflow of silt. The whole or a part of a seam may have been removed by river erosion during the period of accumulation, the channel course (washout) being occupied by a shoestring sandstone (Section 3.5.4). Of more general importance for the coalfield geologist are the effects of folding and faulting during the late stages of the Hercynian orogeny and of erosion prior to the deposition of the overlying Permian and Mesozoic cover. Most British coalfields are located in broad faulted synclines between which lie uplifted regions from which the productive measures have been removed by erosion (Fig. 3.17). In Belgium and the Ruhr, near the Hercynian orogenic front, coal seams are strongly folded and are disrupted by thrusts. Concealed coalfields in central, southern and eastern England lie unconformably beneath an undisturbed cover of Permian and Mesozoic strata and are worked beneath this cover in Kent, Nottingham, Durham and adjacent parts of the North Sea.

In Britain, individual coal seams ranging up to about 3 m in thickness are mined underground largely by mechanised cutting and loading equipment. Ease of development depends on uniformity of thickness and quality and on absence of splitting, washouts and minor faults. The expense of developing deep mines limits the distance for which seams can be worked down dip and the technical difficulties of adapting machinery to changes of level may make it uneconomic to mine in heavily faulted regions. The characters of the floor and roof rocks affect the safety of the mine. Thin coaly partings in overlying strata may render the roof liable to fall; soft seat earths (Table 3.5) may tend to bulge and distort underground roadways; permeable sandstones, faults or zones of close jointing may channel water into the mine. Methane (CH_4, firedamp) diffusing from the surrounding rocks provides a potential cause of explosions which must be countered by effective ventilation. Although these features may be predictable in a general way, they can only be evaluated on the basis of detailed acquaintance with each individual mine and with the structure recorded during earlier stages of mining.

References

Hobson, G. D. and E. N. Tiratsoo 1981. *Introduction to petroleum geology*, 2nd edn. Palo Alto, California: Scientific Press.

Ion, D. C. 1980. *Availability of world energy resources*, 2nd edn. London: Graham & Trotman.

Open University 1974. *The Earth's physical resources: Energy resources*, S266, Block 2. Milton Keynes: The Open University.

Open University 1976. *Earth science topics and methods*: Sedimentary basin case study, S333. Milton Keynes: The Open University.

Raistrick, A. and C. E. Marshall 1939. *The nature and origin of coal seams*. London: English Universities Press.

Selley, R. C. 1978. *Ancient sedimentary environments and their subsurface diagnosis*. London: Chapman & Hall.

Tomkieff, S. I. 1954. *Coal and bitumen: nomenclature and classification*. Oxford: Pergamon Press.

Trueman, A. E. (ed.) 1954. *The coalfields of Great Britain*. London: Edward Arnold.

Williamson, J. A. 1967. *Coal mining geology*. Oxford: Oxford University Press.

4 Metals and their sources

4.1 Introduction

4.1.1 Ore deposits

Most metals used for industrial purposes occur in the Earth's crust in average amounts measurable only in parts per million. These metals can be mined only where they have been enriched by natural processes, usually to concentrations at least two orders of magnitude above the crustal average; ore deposits are the natural concentrations of metals which repay working. The economic proviso built into this definition takes account of the metallurgical and financial considerations that enter into the assessment of a potential ore body. Although the non-geological aspects of mineral exploration will not be examined in depth, Table 4.1 lists factors that may influence decisions concerning the development of an ore field; others will probably occur to the reader. As we have seen already (Section 1.1, Table 1.3) changes in market price may critically alter the status of sub-economic deposits.

Anomalous concentrations of any element are, in the nature of things, exceptional occurrences, and the study of metalliferous mineral deposits is, therefore, concerned largely with unusual geological events. The discrepancies between average crustal abundances and abundances in representative ore deposits (Table 4.2) demonstrate the extent to which ore-forming processes depart from the geological norm. The special characters of these processes are connected largely with the distinctive properties of the ore-forming minerals.

Table 4.1 Assessment of potential orefields.

Geological factors
principal metals, grade (i.e. concentration) of ore

possible byproducts – (e.g. silver as trace metal enhances value of PbZn deposits)

size and structure of ore body

estimated reserves

Mining and metallurgical considerations
mining procedure necessary, e.g. opencast/underground

extraction processes required (e.g. ease of separation of ore minerals, suitability for cheap treatment of purified ore)

suitability of end product (e.g. low levels of impurities such as P in iron ore desirable for use in steel making)

Other considerations
price of metal on world market, expected future demand

ease of access, cost of development

legislation relating to mining

likelihood of future changes in social, political or economic climate liable to affect development

Table 4.2 Discrepancies in elemental concentration between crustal averages and sample ore bodies.

log scale	Average abundance in crust	Concentration in sample ore bodies
10%		
1%	Al(8.1) Fe(4.7) Mg (1.9)	in Hamersley Range Fe ⩾ 65%
1000 ppm	Ti Mn	modern manganese nodules (Mn) up to 40%
100 ppm	S	
10 ppm	V Cr Zn Ni Cu Co Pb	Kambalda Ni 3.5% porphyry coppers Cu 0.5–1%
1 ppm	U Sn As Mo W	for U see Table 1.3
<1 ppm	Sb Cd Hg Ag Pd Pt Au	Merensky reef, Bushveld Complex, Pt+Pd 5.7 ppm

4.1.2 Mineralogy

Whereas most rock-forming minerals are silicates, the great majority of ore minerals are sulphides, oxides and other non-silicates (Table 4.3). The physical properties and chemical behaviour of these minerals differ from those of the common rock-forming silicates to an extent which facilitates the segregation of ore minerals by natural processes. Metallic sulphides, oxides and native metals, for example, have high relative densities and therefore segregate along with the heavy mineral fraction during sedimentary differentiation (see placers, Section 4.4.2). The principal chemical classes of ore minerals are outlined in Table 4.3. The names and compositions of minerals mentioned in this chapter are tabulated in Appendix 1, which also gives a brief summary of their diagnostic properties.

In most ore deposits, the important metalliferous minerals are intermixed with quartz, calcite or other minerals with little or no economic value. These unwanted minerals constitute the **gangue** which is discarded during mining and processing. The proportion of ore minerals relative to gangue, and the ease with which they can be separated, affect the value of an ore deposit. **Disseminated sulphides**, for example, are scattered through the host rock in small grains and aggregates as in the porphyry coppers (Section 4.3.4). **Massive sulphides** are aggregates almost free from other minerals and they may form rich pockets which are easy and profitable to mine. The gangue minerals, though seldom intrinsically valuable, provide useful evidence concerning the environment of mineralisation and a characteristic mineral suite such as the baryte–fluorite association of the English Pennines may serve to define the extent of an orefield (Section 4.6).

4.1.3 Ore genesis

Metalliferous mineral deposits can be grouped in relation to the geological processes by which they are formed or in relation to the metals they contain. The grouping adopted here is a geological one (Table 4.4) and Sections 4.3 to 4.6 deal with the main types of deposit according to their geological associations. Table 4.5 provides for reference a list of the principal metals grouped accord-

Table 4.3 Ore minerals: an outline classification (see Appendix 1 for mineral names, compositions and properties).

Sulphides (and related compounds) have the general formula A_mX_n where A represents one or more metals and X sulphur, arsenic, antimony, bismuth, selenium or tellurium. Ag, Au, Co, Cu, Fe, Hg, Mn, Ni, Mo, Pb, W and Zn form important sulphide ore minerals.

Sulphosalts have the general formula $A_mB_nS_p$ where A represents one or more metals, B one of the semi-metals As, Bi or Sb and S is sulphur.

Oxides, hydroxides and oxygen salts, among which are ore minerals of Cu, Cr, Fe, Mn, Ti, Sn, W and Zn together with Th and U.

Native elements including Ag, Au, Cu, Pt and the non-metal C (as diamond).

Others including halides, carbonates, nitrates, borates, phosphates, tungstates and silicates.

Table 4.4 Metalliferous mineral deposits: a geological grouping.

(a) *Deposits related to igneous processes*
orthomagmatic: segregated during consolidation
pneumatolytic: associated with residual fluids
exhalative and fumarolic: deposits of volcanic centres
some hydrothermal deposits (see (d))

(b) *Deposits related to sedimentary processes*
placers and related deposits: segregated by physical processes
metalliferous chemical sediments including deep-sea deposits
metalliferous residual sediments and weathering products
some hydrothermal deposits (see (d))

(c) *Deposits related to metamorphic processes*
pyrometasomatic deposits: formed at igneous contacts
other metasomatic deposits
deposits of groups (a), (b) and (d) modified by metamorphism

(d) *Hydrothermal deposits*: formed through the agency of hot waters circulating in the host rocks:
vein deposits
replacement deposits

Table 4.5 Metals and their sources.

	Principal types of deposit	Approximate annual production (1977) (tonnes)
Precious metals		
gold	hydrothermal, placer	1200
silver	volcanic, hydrothermal	70 000
platinum	orthomagmatic	700
Light metals		
aluminium	residual sedimentary hydrothermal	14 million
Steel industry metals		
iron	sedimentary	850 million
nickel	magmatic, residual	500 000
manganese	sedimentary	10 million
chromium	orthomagmatic	9 million
cobalt	sedimentary	20 000
molybdenum	hydrothermal, volcanic	70 000
tungsten	pneumatolytic	45 000
Non-ferrous metals		
copper	volcanic, sedimentary	6 million
tin	pneumatolytic, placer	200 000
zinc	volcanic, hydrothermal, sedimentary	4 million
lead	volcanic, hydrothermal, sedimentary	3.5 million

Production based on *Mining Annual Review*; estimates (usually exclusive of Eastern Europe and China) give order of magnitude only.

ing to their industrial uses and a summary of their principal sources.

It is customary to draw a distinction between **syngenetic** ores, coeval with the rocks in which they lie, and **epigenetic** ores deposited after the formation of the enclosing rocks. In broad terms, ores formed by sedimentary and orthomagmatic processes are usually syngenetic, whereas pneumatolytic and hydrothermal ores are epigenetic (Table 4.4). For the reasons outlined below, the boundaries between the two classes are somewhat unreal.

Although a single act of mineralisation sometimes leads directly to the formation of a workable ore deposit (such as the orthomagmatic chromite seams described in Section 4.3.2), mineralisation more often seems to have been achieved in stages, a fact hardly surprising in view of the low average concentrations of many metals in the Earth's crust (Table 4.2). Two examples will illustrate the point. The banded iron formations of the Hamersley Range, Western Australia, were enriched in iron by chemical–organic processes during sedimentation and are therefore syngenetic (Section 4.4.4). The average iron content is below 55%, but the redistribution of iron after a subsequent phase of folding and fracturing has locally raised concentrations to 65%; and the principal mines are sited in the areas of epigenetic enrichment (Fig. 4.1). Secondly, tin mined on a small scale in central Swaziland is derived from the weathered mantle developed over pegmatites carrying cassiterite (SnO_2). Tin values in the unweathered pegmatites are below economic levels, but the partial removal of weathered feldspar and other silicates by hill creep and other surface processes has effected a further concentration of cassiterite which renders small-scale recovery economic.

The final stage in the processes leading to mineralisation often involves the intervention of hot, chemically active pore fluids. The solubility and stability of many ore minerals vary according to redox conditions and pH; sulphides, for example, are easily oxidised in aerobic environments. **Hydrothermal** deposits, precipitated from migrating solutions as replacements of the host rock or as veins filling open fractures, form an important class of ores which may be associated with many of the other groups shown in Table 4.4.

The extent to which two or more processes contribute to produce economic concentrations of metals increases the number of factors influencing the distribution and structure of the resulting deposits. Ore bodies of a single genetic type may vary widely

51

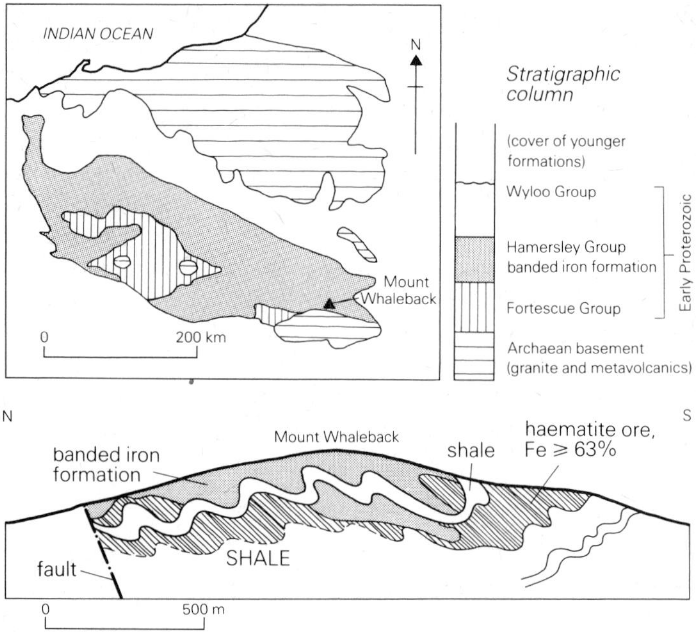

Figure 4.1 The formation of an ore body Two stages in the concentration of metals during formation, illustrated by banded iron formations, Hamersley Range, Western Australia. The stratiform banded iron formation (see Section 4.4.4) is a widespread sedimentary deposit (above). At Mount Whaleback and other localities, the grade of ore has been raised by post-depositional processes to percentages that repay mining (below). The shale band acts as a marker horizon, showing that the high-grade ore represents an enriched part of the BIF. (Section based on Kneeshaw, M. 1976. *25th International Geological Congress Excursion Guide* 50.)

among themselves and the location of major deposits within a single province often seems to be unpredictable – a freakishness expressed by the prospector's maxim 'gold is where you find it'. It is impossible to deal with all the contributory factors in a book of this kind, but it should be emphasised that the local details determining the siting of individual deposits are of crucial importance to the prospector and the mine geologist.

4.2 Mineral exploration

The opening of a new mine is commonly the end result of many years of investigation aimed, first, at identifying a suitable deposit and, secondly, at establishing its potential value (Table 4.6). The costs of exploration are high, amounting to 1% of the value of the metals discovered in North America over the period 1950–70. The techniques mentioned in this section are explained and illustrated in Chapter 8.

Reconnaissance methods designed to identify favourable anomalies must be cap-able of covering large areas at relatively low cost. The target areas indicated by reconnaissance or by random finds (many ore fields have been discovered accidentally or by the efforts of prospectors using extremely simple methods) are subjected to more detailed examination designed to locate individual ore bodies at or below the surface and to establish the factors controlling the distribution of ore.

The search for new deposits in an established orefield takes account of the geological setting and structural relationships of the known ore bodies; for example, the buried extensions of stratiform ore bodies such as the Witwatersrand gold reefs (Section 4.4.2) may be predictable when their places in the stratigraphical succession and the regional structure are known. More specific techniques exploit the distinctive properties of metalliferous ore minerals –high density, high conductivity and strong magnetic signatures or (for uranium and thorium) radioactivity. The rusty red–brown oxidised crust or **gossan** formed by weathering of sulphides may pinpoint a deposit.

Table 4.6 Exploration procedures: an idealised sequence (for details of techniques see Ch. 8).

Decision to prospect, based on:
regional geology favours mineralisation of desired type

evidence of mining in former times

random finds of gossan or 'shows' of sub-economic
    mineralisation

Regional reconnaissance by:
airborne geophysical surveys (magnetic, electromagnetic,
    radiometric, gravimetric)

geochemical reconnaissance (stream-sediment or
    lake-sediment sampling)

photogeological survey, including use of satellite imagery

ground geological reconnaissance, e.g. traverse mapping

    leading to:
    identification of favourable anomalies, staking of claims;
       random finds, staking of claims; or
    rejection of unfavourable areas, abandonment of project

Investigation of selected target areas:
geological mapping

ground geophysical surveys (gravimeter, magnetometer,
    resistivity and induced potential surveys)

detailed geochemical surveys (closely spaced
    stream-sediment sampling, sampling of drift, soil)

exploratory pitting, trenching, trial boreholes

    leading to:
    discovery of deposits, staking of claim;
    identification of probable buried ore body; or
    abandonment of project

Assessment of ore body
detailed topographic and geological survey of site

further boreholes, logging of cores

petrographical and chemical study of cores

assaying of ore samples

    leading to:
    decision to develop;
    decision to suspend operations; or
    decision to relinquish claim

Mineralised rock fragments in glacial drift or stream sediments, or anomalous metal values in the fine fraction of stream sediments, may indicate an ore body up stream. Shallow pitting and trenching may be used to penetrate a drift or soil blanket or to investigate 'shows' or gossans.

The discovery of a potential ore body marks the start of a precise investigation of its three-dimensional structure and geochemistry. Surface mapping is supplemented at this stage by extensive drilling, logging of the core and assaying of ore samples, with a view to establishing the grade and probable extent of the ore. The method of mining to be adopted depends on the size, structure and type of ore body. Small ore bodies of high value such as gold-bearing quartz veins may be mined out from open pits or shallow shafts. Large low-grade ore bodies such as porphyry coppers are usually mined opencast, as are iron ores which yield a metal of relatively low value (Fig. 4.2). Placers or residual concentrations in the weathered zone may be dredged or dug from stream, beach or land surface. Deep mining with its attendant expenses and technical problems is reserved mainly for ore bodies of proved extent and high intrinsic value such as the Witwatersrand gold–uranium reefs (Section 4.4.2).

4.3 Ore deposits related to igneous activity

4.3.1 Magmatism and associated processes
Magmas produced by partial melting in the mantle or lower crust contain metals which may be segregated within the igneous rocks formed on consolidation to give magmatic ore deposits of various types. Heat from rising magmas may set in motion hydrothermal circulation systems capable of redistributing any metals already present in the country rocks. The principal types of ore deposits directly or indirectly related to magmatism are summarised in Table 4.7, which introduces some new terms (see also Table 4.4).

Orthomagmatic deposits are those formed during the main period of consolidation, usually in intrusive complexes. Metals deposited from the active, volatile-enriched residuum of the magma at a late stage of consolidation are said to form **pneumatoly-**

Figure 4.2 Opencast mining   The mining of iron ore at Hibbing, Minnesota (Aerofilms).

tic deposits; they occur in minor intrusions, veins and replacement bodies formed by reaction between volatiles and previously formed rocks. **Pyrometasomatic** deposits are formed by reaction with the adjacent country rocks.

Fluids escaping from submarine volcanic vents may give **exhalative** deposits formed on the sea floor and volcanic gases escaping from terrestrial vents may yield **fumarolic** deposits. Hydrothermal activity involving waters heated by rising magma, and often partly of juvenile origin, leads to the formation of vein systems and replacement bodies which may be zonally arranged around plutonic or subvolcanic centres.

The extent and style of mineralisation associated with any igneous province depend on the composition of the parent magmas and on the manner in which differentiation has taken plac during consolidation. In so far as these variables are related to the tectonic setting, a fivefold division can be made:

Table 4.7 Ore deposits of igneous affinities.

Igneous assemblage & tectonic setting	Dominant ore-forming processes	Forms and characters of deposits	Examples	Principal metals
(a) mid-ocean ridge volcanics (constructive plate margins, ophiolites)	(i) exhalative and hydrothermal in lavas	stratabound massive sulphides formed on sea floor, or veins and replacement bodies	Troodos Complex, Cyprus	Cu(Ni)
	(ii) orthomagmatic in ultrabasic intrusions	layers or irregular bodies	Ural Mountains	Cr
(b) greenstone belts (early Precambrian, mainly basic volcanics)	(i) exhalative	massive or disseminated sulphides, stratiform iron formations	Abitibi belt, Canada	Cu Zn Fe Au
	(ii) orthomagmatic	chromite layers or sulphide segregations with Ni	Kambalda, W. Australia (Ni)	Ni Cu Cr
	(iii) hydrothermal	quartz veins, carrying gold scavenged from volcanics by hydrothermal fluids	Yilgarn block, W. Australia	Au
(c) island arcs and orogenic mountain belts	(i) exhalative Kuroko type	stratabound sulphides in volcanics usually near rhyolites	Kuroko, Japan	Cu(Zn–Pb)
	(ii) hydrothermal processes round subvolcanic acid stocks – porphyry coppers	disseminated low-grade sulphides in brecciated roof of stock	Bingham, Utah	Cu(Mo)
	(iii) hydrothermal processes mainly in subvolcanic zone	polymetallic sulphides, disseminated and in veins	western USA	Cu Zn Pb Au Ag Sb Mo etc.
	(iv) pneumatolytic and hydrothermal in and round acid stocks and subvolcanic intrusions (mainly of Phanerozoic age)	cassiterite-bearing quartz veins and pegmatite, disseminated ores sometimes associated with tourmalinisation; for uranium see Section 3.4	Bolivia, Andes	Sn (W Nb) U
(d) continental rift valleys	(i) hydrothermal in plateau basalts	native copper or sulphides in vesicles or permeable horizons	Keweenawan lavas, Lake Superior	Cu
	(ii) orthomagmatic in basic–ultrabasic intrusions	stratiform chromite layers or segregations of sulphides	Duluth Complex, Lake Superior	Cr Pt Ni Cu
	(iii) metasomatic in and around carbonatites or alkali complexes		East African rift	Nb, REE
(e) oceanic islands, intraplate oceanic	little important mineralisation			
(f) continental cratons	(i) as for (d)(i)		Siberian lavas	Cu Ni
	(ii) as for (d)(ii)		Bushveld Complex, South Africa	Cr Pt Ti Fe
	(iii) orthomagmatic in anorthosites (mainly of Proterozoic age)	stratiform layers of ilmenite–magnetite	Bushveld Complex, South Africa	Fe Ti
	(iv) pneumatolytic around granite stocks	vein deposits carrying cassiterite	'younger granites' of Nigeria	Sn Nb
	for kimberlite see Ch. 5, diamond			

(a) Constructive plate margins (mid-oceanic ridges) – mainly basic and ultrabasic igneous assemblages.

(b) Destructive plate margins (orogenic belts, volcanic island arcs) – basic and calc-alkaline assemblages including granites.

(c) Continental rift valleys – basic and alkaline assemblages.

(d) Continental cratons (stable interior parts of plates) – basic, alkaline, rare acid assemblages, kimberlites.

(e) Early Precambrian greenstone belts, which have no precise analogues in later environments – basic assemblages with ultrabasic and acid associates.

As Table 4.7 shows, certain metals are preferentially associated with igneous rocks of a particular composition. Chromium, nickel and the platinum metals are almost invariably associated with basic or ultrabasic rocks. Copper, zinc and gold frequently concentrate in the intermediate to acid differentiates associated with basic volcanic provinces. Copper, zinc, lead, gold, silver and molybdenum are associated with the varied calc-alkaline volcanics of island arcs and orogenic belts, whereas tin, tungsten and uranium are linked with acid plutonic and subvolcanic centres. Uranium, thorium and rare-earth elements characterise alkaline complexes whereas diamonds are virtually confined to kimberlites (see Section 5.7).

4.3.2 Orthomagmatic ore deposits

Economic ore deposits of chromium, nickel or the platinum metals are formed in ultrabasic–basic intrusions (or occasionally in lavas) by **magmatic differentiation**. The crystallisation of chromite (Cr_2O_3), the only ore mineral of chromium, is controlled by magma composition and oxygen concentration. Under some conditions, a critical stage is reached at which the mineral crystallises in large amounts and, being dense, piles up on the floor of the magma chamber. Subhorizontal chromite-rich bands are thus interleaved with layers formed by sinking of the early-formed silicates olivine, pyroxenes or plagioclase. Platinum and allied metals appear in trace amounts and, in appropriate chemical environments, magnetite and ilmenite may also form subhorizontal layers.

All the metals mentioned above are mined in the Bushveld Complex of South Africa, one of the world's most richly mineralised intrusions (Fig. 4.3). The Precambrian Bushveld Complex forms a group of gigantic sagging sheets (**lopoliths**) in which banded gabbros, norites and anorthosites reach maximum thicknesses of 8 km. The subhorizontal layering of the basic igneous rocks is ascribed to rhythmic deposition of silicates and oxides from the cooling magma. Diorites and granites enriched in Fe, Si, alkalies and other incompatible elements represent the last residuum of this magma. They are overlain by a 2 km sheet of intrusive granite, possibly derived from melting of the crust by the hot basic magma.

The ore deposits of the complex yield chromium, platinoids, iron and vanadium from the layered cumulates and tin and its associates from late phases of the granite suite. The deposits of the layered sequence are as a rule stratiform, and are located at consistent levels in the igneous succession (Fig. 4.3).

The concentration of nickel in igneous complexes is due in part to the fact that sulphides have a limited solubility in silicate magmas. At certain concentrations of sulphur, a sulphide melt carrying nickel, copper and iron separates as droplets which may unite to form a body of dense fluid. Massive sulphide bodies carrying pyrrhotite, pentlandite and chalcopyrite (see Appendix 1) may be formed from this fluid, often near the base of the intrusion or as vein-like bodies in the underlying rocks. By far the largest source of nickel is the Precambrian Sudbury Intrusive of Ontario, a dish-shaped igneous sheet passing from mafic norite at

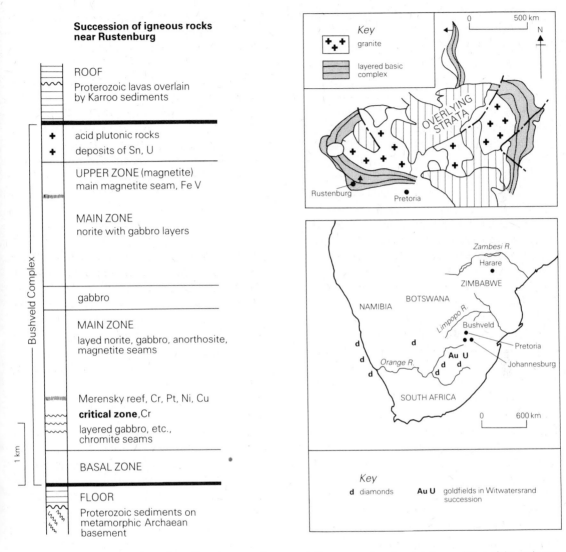

Figure 4.3 The Bushveld Complex Sketch map and representative succession showing the positions of the main ore deposits. The locality map (bottom right) gives the location of deposits of gold and uranium and diamonds in southern Africa.

the base to acid rock at the top. Nickel-bearing massive sulphides occur sporadically around the periphery of the basin where the igneous body overlies its country rocks.

4.3.3 Pneumatolytic and pyrometasomatic deposits

The residual magmatic fluids which remain when the bulk of an intrusive magma has consolidated are enriched in incompatible elements such as K, Rb, Be, Li, U and Th which do not readily enter the lattices of high-temperature silicates, and in volatile components such as H_2O, CO_2, F, Cl, P and S. Cu, Fe, Sn, W and Nb can form complex radicles with these volatiles and may be deposited by pneumatolytic action in dykes, veins and replacement bodies. Such activity, usually at 550–600 °C, is most common in

57

Pegmatites, the common derivatives of late-stage magmatic fluids, are characterised by an exceptionally coarse grain size, mineral growth being aided by the activity of volatiles. Granite pegmatites consist mainly of alkali feldspars and quartz, with muscovite as a common accessory. Mineralised granite pegmatites may also carry tourmaline, topaz, beryl or lithium mica as well as cassiterite (SnO_2, the only ore mineral of tin), tungsten minerals, pitchblende (Section 3.4.2), thorianite (ThO_2) or the cerium-bearing phosphate monazite. Apatite and minerals carrying uranium, thorium or rare-earth elements occur in pegmatites related to alkaline complexes. The economic importance of pegmatites arises not only from their contents of rare elements but also from the occurrence of gemstones and of coarsely crystalline feldspar and mica used in ceramics and for insulation respectively (Ch. 5).

Reaction of volatiles with previously consolidated igneous rocks or with the adjacent wall rocks may lead to replacement of feldspar by white micas and/or tourmaline, a form of alteration which is sometimes associated in granites with the deposition of cassiterite. Spectacular alteration products may be formed in limestones, especially where volatiles are rich in iron. **Skarns** are coarsely crystalline aggregates of iron and calcium silicates in marble which may be associated with economic concentrations of magnetite, haematite and sulphides. The term **pyrometasomatism** is sometimes used for the high-temperature reactions by which skarns are formed.

Classic examples of pneumatolytic deposits are the orefields centred on the Hercynian granites of south-west England which have been worked since prehistoric times (cf. Fig. 1.3). The exposed granites represent high points or cupolas on a buried batholith and appear to have provided foci for pneumatolysis and the generation of convective hydrothermal circulation systems (Fig.

4.4). Pegmatite dykes and sills invading the adjacent Palaeozoic slates (locally known as killas) pass, away from the granites, into hydrothermal quartz veins. White mica and tourmaline are characteristic minerals, not only in pegmatites but also in metasomatic replacements in both granite and killas. Greisen, consisting almost entirely of quartz, white mica and topaz, and quartz–tourmaline rocks are end products of replacement. Deposits of tin (as cassiterite) and tungsten (as scheelite) are concentrated in an inner zone enclosing the granite outcrops. An outer zone, five or more kilometres in breadth, is characterised by hydrothermal sulphide deposits in which copper and zinc predominate nearer the granites, and zinc and lead at greater distances.

Pitchblende and other uranium minerals are sporadically distributed in the hydrothermal zone. Yet another product of pneumatolysis or hydrothermal action is china clay formed by leaching of silica and alkalies (Section 5.3).

4.3.4 Exhalative, fumarolic and subvolcanic deposits

Ore deposits related to volcanic activity may be located in lavas, in interbedded pyroclastic or sedimentary rocks, in high-level intrusive bodies, around the conduits which transmit magma or at the vents by which these conduits discharge. The Tertiary volcanic provinces marking the destructive plate boundary of Western North America, Central America and the Andes carry a rich and varied suite of ores (Table 4.7c) and, at the other end of the timescale, the early Precambrian greenstone belts of Canada, South Africa and Western Australia are almost equally productive (Table 4.7b).

The discharge of metalliferous fluids from submarine volcanic vents may lead locally – as in the deep parts of the Red Sea – to the accumulation of pools of dense metalliferous brine in topographical hollows (for composition of brine see Table 2.1). Metalliferous

58

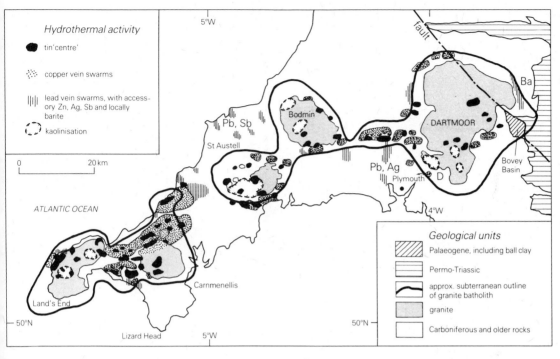

Figure 4.4   The south-west England orefield   The distribution of pneumatolytic tin deposits, hydrothermal deposits of Cu, Pb, Zn, Ag, etc. and of china clays and ball clay (Ch. 5) is shown in relation to Hercynian granites (modified from Moore, J. McM. 1982. *Metallization associated with acid magmatism.* New York: Wiley).

sediments precipitated from such modified sea water include bedded cherts mixed with iron or manganese minerals and, locally, Fe-, Cu- or Zn-bearing sulphide bodies. The Atlantis II Deep of the Red Sea is estimated to contain 2.5 million tonnes of Zn. **Exhalative** sulphide deposits are often lenticular in cross section, and are commonly located above a vertical pipe-like mineralised zone marking the channel through which fluids reached the sea floor. Exhalative sulphide deposits of island arcs include the Kuroko-type copper deposits of Japan.

In sub-aerial volcanoes, volcanic gases may deposit encrustations of sulphur and sulphides carrying mercury, antimony and other metals around gas vents or **fumaroles**. In the nature of things, these deposits are vulnerable to solution and erosion and are therefore preserved mainly in young volcanic provinces.

Ore deposits of subvolcanic environments include the important porphyry coppers of many Mesozoic and Tertiary orogenic belts and island arcs characterised by calc-alkaline volcanicity (Fig. 4.5). Mineralisation of this type is centred around small (usually <3 km diameter) porphyritic granodioritic to monzonitic stocks emplaced at depths of only a few kilometres. Ores containing pyrite and chalcopyrite (sometimes accompanied by molybdenite) are concentrated in the uppermost parts of the stock and commonly extend into the enveloping wall rocks. As the porphyry copper of Bingham, Utah, illustrates, a concentric zoning is often apparent, an inner zone of disseminated Cu ± Mo sulphides being enclosed by a zone in which the same sulphides occupy a fine network of veinlets and, sometimes, by a zone characterised by Pb, Zn and Ag (Fig. 4.5). The host rocks in and even beyond the mineralised area show in-

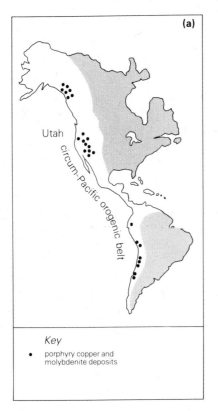

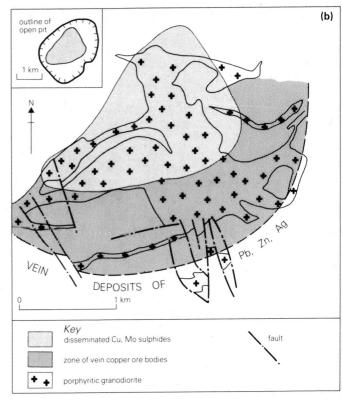

Figure 4.5 Porphyry copper deposits   (a) The principal porphyry copper and molybdenite deposits of the Americas in relation to the circum-Pacific orogenic belt. (b) Mineralisation centred on an irregular porphyritic granodiorite at Bingham Canyon, Utah: the inset shows the size of the pit from which ore is extracted. (Simplified from Dixon, C. J. 1979. *Atlas of economic deposits.* London: Chapman and Hall.)

tense alteration to chlorite and muscovite ± calcite ± epidote, or to clay minerals.

Mineralisation is thought to result from the rapid loss of sulphur-charged volatiles as the rising magma nears the surface. The exsolved volatiles escaping upwards and outwards generate pressures sufficient to shatter the rocks through which they pass and deposition in the fractures so formed produces the outer vein network; the alteration halo records the effects of reaction between volatiles and country rocks. The low grade (often ~1% Cu) and the disseminated style of mineralisation make it necessary to mine the entire rock opencast. The vast tonnages available and the possibility of recovering small amounts of silver or gold give the deposits great economic importance.

4.4   Ore deposits related to sedimentation, diagenesis and weathering

4.4.1   *Mineralisation in sedimentary environments*
Sedimentary ore deposits share many of the characteristic structural features of sedimentary rocks in general (Table 4.8). Ore bodies concordant with bedding are said to be **stratiform**. Most such bodies are sheet-like, but placers and other detrital accumulations

Table 4.8   Ore deposits formed by surface processes.

Host rocks and environment	Dominant ore-forming processes	Forms and characters of deposits	Types of deposit and examples	Metals
(a) rocks in zone of weathering and residual sediments	(i) concentration *in situ* by removal of soluble components, especially in tropical zones	irregular layers and patches at surface or below unconformities	bauxite, Jamaica; silicate nickel, New Caledonia; laterite, tropical Africa	Al Ni Fe and as building material
	(ii) solution in oxidised zone deposition near water table	zones of secondary enrichment of sulphide ore bodies at surface or below unconformities		Cu Zn Pb
(b) detrital sediments	(i) mechanical segregation of heavy minerals	in disaggregated weathered mantle (eluvial), as concordant layers or lenses in fluviatile and beach deposits	placers: Orange River, Namibia (diamonds); Queensland coast (monazite); California (gold)	Au Sn diamond monazite
	(ii) as (b)(i), but heavy mineral concentrations diagenetically reworked	ancient placers represented by pebble conglomerates or quartzites	Witwatersrand (South Africa); Blind River, Ontario	Au U U
	(iii)	stratiform, lensoid	Sullivan, British Columbia, Broken Hill, New South Wales	Pb Zn
	(iv) deposition from pore fluids during or after diagenesis, sometimes under reducing conditions	lensoid and irregular bodies	'red-bed' copper deposits, Colorado plateau; roll-front uranium (see Section 3.4)	Cu V U
(c) argillaceous sediments	adsorption on clay particles, reduction of sulphate by bacterial action	usually associated with carbonaceous pelites	Kupferschiefer (Permian) of Germany	Cu (Zn–Pb)
(d) carbonate rocks and evaporites	deposition from circulating saline waters	irregular replacement bodies, veins, breccia fillings, often concentrated in limestone reefs, dolomites or evaporitic horizons	Mississippi Valley type ores	Pb Zn
(e) mixed host rocks	(i) facies control near algal reefs	irregular stratiform	Copper Belt, central Africa	Cu Co
	(ii) acid volcanicity in marine basin	lensoid	Mount Isa, Queensland	Pb Zn Cu
(f) chemical sediments	(i) precipitation and diagenetic reworking, aided by organic activity	banded iron formations: deposits with chert and/ or carbonate layers: entirely Precambrian	Lake Superior, Canada; Hamersley Range, W. Australia	Fe (mainly as oxides)

continued overleaf

61

Table 4.8 Ore deposits formed by surface processes – *continued*.

Host rocks and environment	Dominant ore-forming processes	Forms and characters of deposits	Types of deposit and examples	Metals
(f) Chemical sediments – *continued*	(ii) precipitation in nearshore environments, diagenetic reworking	ironstones, bedded deposits, often oolite	Clinton iron ore, USA; Northampton-shire Jurassic ore, UK	Fe (mainly carbonates and silicates)
	(iii) precipitation in shallow, partly enclosed sea	concordant lenses interbedded with detrital sediments and limestones	Nikopol type, north of Black Sea	Mn (oxides, carbonates)
	(iv) precipitation in deep ocean basins	nodular bodies and encrustations on deep-sea floor	manganese nodules, Pacific Ocean, especially 0–20 °N of Equator	Mn (Fe) Ni
	(v) exhalative metalliferous sediments; see Section 4.3.4			

formed in stream channels may have a ribbon-like form. The distribution of ores (such as those of the Copper Belt, Section 4.4.2) that relate to a particular environment of deposition reflects broad palaeogeographical and facies controls. In detail, however, the structure and texture of many deposits record modifications due to diagenetic and/or low-temperature hydrothermal processes acting in the post-depositional period. Nodular segregations of sulphides – especially where remnants of organic matter provide centres of reduction – and crystalline encrustations on joints, fractures and cavities may be formed by these processes. Migrating pore fluids carry metals scavenged from the sediment to new sites where they are concentrated in new chemical environments. Ore bodies resulting from reworking within a sedimentary succession are not concordant in detail, though many remain **stratabound**, that is, confined within a single stratigraphical unit.

4.4.2 Deposits in detrital sediments (Table 4.8b)

Placer deposits, formed by the accumulation of heavy minerals derived from the erosion of older mineralised terrains, appear in alluvial formations and beach sediments where mechanical sorting is assisted by current or wave action. The majority of known placers are geologically young; indeed, many are related to existing river basins and shorelines (cf. diamond placers, Section 5.7). Alluvial gold derived from Tertiary hydrothermal veins was a principal target of the 19th- and early 20th-century gold rushes to California and the Klondike, the gold being separated from stream sediment by panning (Section 8.3). Alluvial and marine cassiterite placers supply most of the tin production of Malaysia and Thailand, respectively, where the parental tin-granites are of late Palaeozoic and early Mesozoic age. Many of these unconsolidated deposits can be dredged or scooped from the river bed.

Older placers occur sporadically in successions of suitable facies but (because terrestrial and coastal sediments are vulnerable to erosion) they are less numerous than those mentioned above. One group of Precambrian gold and/or uranium deposits that appears to have originated as placers is, however, of overwhelming importance. This is the Witwatersrand conglomerate type which is represented in the type example of South Africa (Au, U), the Jacobina deposits of Brazil (Au), the Blind River deposits north of Lake Huron (U) and perhaps in Northern Australia (U). This type of modified placer has been mentioned in connection with uranium (Section 3.4).

In the Johannesburg region of South Africa, an Archaean basement containing hydrothermal gold deposits is overlain unconformably by a thick succession of which the Witwatersrand system deposited about 2600 million years ago, is the most important unit (Fig. 4.6; for location see Fig. 4.3). The succession, of interbedded sandstones, conglomerates and shales, occupied a basin at least 350 km by 200 km which was partly enclosed by highlands undergoing erosion. Cross bedding and other palaeocurrent indicators show that streams carrying detritus entered this basin from several points at its northern and northwestern margin (Fig. 4.6). Mineralisation is stratabound, being concentrated in conglomerate layers, usually only a few metres in thickness. The conglomerates, known locally as reefs or bankets, are best developed near the northern basin margin, where richly mineralised 'pay streaks' representing channel fillings radiate from the entry points of the river systems. The reefs crop out in the city of Johannesburg, the site of the original finds; the deep mines worked today are clustered above their southward-dipping extensions.

The conglomerates are made up of close-packed small quartz pebbles in a matrix containing detrital quartz, zircon and chromite, together with pyrite, uraninite and minute gold particles. Much of the uranium is

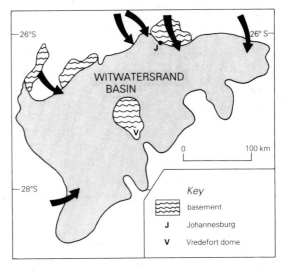

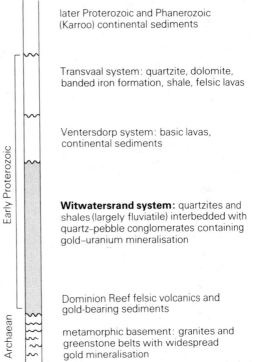

Figure 4.6 The Witwatersrand Group The sketch map (top) shows the probable original extent of the basin and the entry points or rivers supplying detritus to it (arrows). Over much of the basin, the Witwatersrand strata are buried beneath Ventersdorp, Transvaal or younger rocks. Deep mines penetrate this cover to reach the gold-bearing horizons. The stratigraphical column shows the position of the Witwatersrand and locally developed Dominion Reef sediments above a gold-bearing basement.

63

concentrated in veinlets and much of the pyrite (which forms up to 10% of the rock) forms fine-grained mosaics which evidently crystallised *in situ*. These features have been held to indicate that the ores are not sedimentary but epigenetic hydrothermal deposits localised in conglomerates because these rocks were permeable to mineralising solutions. An alternative favoured by many geologists at present is that the ore bodies originated as placers but were later modified by diagenetic and/or hydrothermal processes.

Other ore deposits in clastic host rocks have been mentioned in connection with roll-front uranium ores (Section 3.4). Analogous red-bed copper and vanadium ores are formed by reaction with metalliferous pore fluids, especially where organic matter provides reduction centres (Table 4.8b, (iv)). Many argillaceous sediments, especially those with a high organic content, carry base metals adsorbed on clay particles and sulphides formed by reduction of sulphate in sea water. A deposit of this type is the Permian Kupferschiefer of Germany and Poland, a bituminous dolomitic shale accumulated in a saline anaerobic basin, which yields copper and lesser amounts of lead and zinc (see Fig. 5.3 for succession).

Finally, the Copper Belt of central Africa

may be mentioned here, although the host rocks include both detrital and carbonate sediments (Fig. 4.7; Table 4.8e). The late Precambrian Katangan Supergroup rests unconformably on a basement of granites and schists which formed a land surface marked by low hills of granite. The lowermost Katangan formations, largely confined to the original hollows, are terrestrial. A change in the style of deposition associated with copper mineralisation followed an incursion of the sea. The growth of algal reefs around remaining topographical 'highs' led to the development of lobate bodies of dolomitic limestone between which black siltstones and shales accumulated in partly enclosed gulfs. The ore bodies of pyrite, chalcopyrite, bornite and chalcocite (see Appendix 1) are located in shaly units of the Ore Formation (Fig. 4.7) and are closely controlled by sedimentary facies. They have, however, been modified by solution and redeposition during a period of folding and low-grade metamorphism which affected the Copper Belt near the end of the Precambrian era.

4.4.3 Deposits in carbonate rocks
       (Table 4.8d)

Many limestones, especially reef limestones, contain original voids, or develop voids dur-

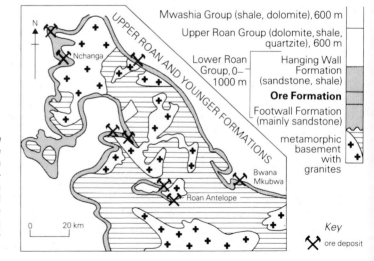

Figure 4.7 Copper deposits in the Katangan Copper belt, Zambia The position of the ore-bearing Lower Roan Group is shown in relation to the basement. The ore formation, 1–65 m in thickness, includes dolomites derived from algal reefs, and copper sulphides are concentrated in the dark shales fringing these reefs (based on various compilations of F. Mendelsohn).

Mwashia Group (shale, dolomite), 600 m

Upper Roan Group (dolomite, shale, quartzite), 600 m

Lower Roan Group, 0–1000 m

Hanging Wall Formation (sandstone, shale)

Ore Formation

Footwall Formation (mainly sandstone)

metamorphic basement with granites

Key

✗ ore deposit

ing dolomitisation or solution weathering. They therefore provide favourable hosts for diagenetic or hydrothermal ore deposits, as they do for oil and gas (Section 3.5.4). Lead–zinc sulphide ores of Mississippi Valley type are stratabound deposits in limestone hosts which are among the principal sources of Pb and Zn in the USA. The type examples around the junctions of the states of Missouri, Tennessee and Oklahoma are located in Cambrian to Carboniferous limestones, usually of shallow-water reef facies. The connection with this sedimentary facies is also well illustrated by the Pine Point deposit near Great Slave Lake, Canada (Fig. 3.7) which is located in a Devonian reef.

The principal ore minerals (sphalerite (ZnS) and galena (PbS)) and the common gangue minerals (fluorite and baryte) locally form bedded concentrations but more often appear as irregular replacements and as crystalline vein fillings and encrustations on joints and voids. The crustified growth of minerals layer on layer indicates that deposition took place in many phases (cf. Fig. 4.10). The composition of fluid inclusions trapped in the sulphide or gangue minerals suggests deposition from saline water at temperatures of 100–150 °C; the isotopic composition of the lead suggests that these brines carried metals derived from more than one source. Traces of hydrocarbons are found in many deposits and may have assisted in their formation.

Mineralisation of Mississippi Valley type is usually ascribed to brines expelled during the diagenesis of thick successions. These connate waters, heated by deep burial, scavenge metals from the sediments traversed in their passage up slope to the basin margin or to interior high points. Deposition of sulphides may be triggered by changes in temperature and composition as the brines enter pervious limestones. More complex factors enter into the formation of hydrothermal deposits such as those of the Pennine district of England (Section 4.6).

4.4.4 Iron ores
(Table 4.8f)

Over 90% of the total of 800–900 million tonnes of iron ore mined annually comes from deposits of sedimentary origin. The formation of these deposits (and of some rather similar manganese deposits, Table 4.8) depends largely on the fact that ferrous iron (Fe^{2+}) and ferric iron (Fe^{3+}) behave differently under atmospheric conditions. Ferric compounds are relatively insoluble in surface waters and are stable in the presence of oxygen. Iron released during weathering may therefore accumulate in ferric compounds with the residual sediments (Section 4.4.6) or in the pore spaces of alluvial sediments. The warm red, brown or yellow colours of haematite (Fe_2O_3) and the hydrated ferric oxide limonite are responsible for the coloration of laterites, gossans and red-bed sediments of terrestrial facies. Transport of iron in solution takes place mainly in the ferrous state and the metal is readily precipitated by changes of Eh or pH, or by mixing of river water with the sea. The stability of ferrous iron compounds was probably greater in early Precambrian times when, as a result of the relatively low level of organic activity, the Earth's atmosphere was poor in oxygen. It is therefore not surprising that sedimentary iron ores fall into two chemically distinct groups: the banded iron formations which are early Precambrian (Archaean to early Proterozoic) in age, and the ironstones which are mainly Phanerozoic (Fig. 4.8).

The **banded iron formations** are the world's most important sources of ore, the principal producers being the USA (Lake Superior region, Fig. 4.2), Canada (Labrador), the USSR (Ukraine), Brazil and Western Australia (Hamersley Range, Fig. 4.1). In all these regions iron formations make stratiform units up to several hundred metres in thickness which show remarkable lateral consistency over distances up to 150 km or more. As their name implies, they show conspicuous alternations of pale siliceous and dark

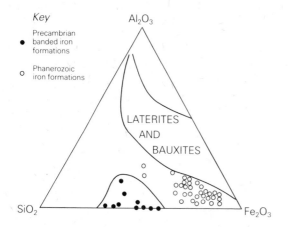

Figure 4.8 Sedimentary iron ores Their chemical composition in relation to the type of deposit. The Precambrian banded iron formations, the principal resources today, contain very little alumina when compared with the Phanerozoic ironstones. Laterites and bauxites formed by chemical weathering in subtropical climates are enriched in alumina and iron as a result of leaching of silica.

ferruginous layers on a scale of a few centimetres. Chert and iron oxides (magnetite and/or haematite) are the principal components, though iron carbonate (siderite) and sulphide (pyrite) are locally important. Deposition appears to have taken place in shallow, tectonically stable basins to which little or no detritus was being added. Precipitation of silica and iron may have been facilitated by organic activity – indeed some authorities consider that oxidation of iron at the site of deposition was due mainly to bacterial action. Complexities of the layering suggest that the observed textures are largely diagenetic. Iron oxides form 40–60% of many banded iron formations which are mined opencast from vast pits (Fig. 4.2). Higher concentrations (~65% Fe) are attained where subsequent haematite enrichment has taken place (Fig. 4.1).

The **Phanerozoic ironstones**, though developed on a smaller scale, are historically important as the principal iron ores mined in Europe and North America during the Industrial Revolution. They include the minette-type Jurassic deposits of Alsace–Lorraine near the German border in eastern France, the Jurassic ironstones of Northamptonshire and other localities in Britain, also Jurassic, and the Silurian Clinton iron ores of eastern USA. Individual ore formations seldom extend continuously for more than 20 km and are usually less than 15 m in thickness; they are commonly oolitic and they pass laterally or vertically into shallow-water sandstones and limestones and represent inshore deposits of sheltered waters. Most contain the iron silicate chamosite with varying amounts of haematite, siderite and other silicates and hydroxides. They carry an appreciable content of Al_2O_3 and have a lower Si : Fe ratio than the banded iron formations (Fig. 4.8).

4.4.5 Deposits of deep oceans

The deep oceans, which lie beyond the reach of terrigenous (land-derived) detritus, accumulate siliceous or calcareous oozes where organic productivity is high. Elsewhere, very slow deposition allows time for reaction between sea water and the volcanic rocks of the sea floor – a reaction which is most effective where volcanic exhalations have enriched the water in metals. Unconsolidated ferromanganese deposits overlie volcanics on some mid-oceanic ridges (cf. Red Sea brines, Section 4.3.4). Manganese nodules accumulating as rounded bodies, often formed as encrustations on rock fragments, cover wide areas of tropical latitudes, especially in the Pacific Ocean west of central America. Manganese oxides and hydroxides are the principal components of the nodules, but in certain areas nickel, copper and other metals are present in amounts totalling a few per cent which make the nodules potentially worth mining (Section 6.5).

4.4.6 Effects of weathering processes (Table 4.8a)

Sulphides and many other ore minerals are unstable during weathering, where oxidising conditions prevail. The consequent release of metals in the zone of weathering may lead to the formation of new ore deposits on or immediately below mature land surfaces.

Metals that do not readily enter solution under surface conditions tend to concentrate in **residual deposits** which remain after soluble components have been removed. By this means, low-grade ores in the bedrock may be enriched in the weathered mantle and soil. Silicate nickel deposits (such as that of New Caledonia) are those that result from weathering of ultrabasic rocks in which silicates such as olivine are rich in nickel. The selective removal of soluble material raises concentrations from about 0.3% Ni in the parent rock to 1.5–2.5% in the weathered product. Selective removal of feldspar during weathering enriches sub-economic concentrations of tin in the cassiterite pegmatites of Swaziland, as was mentioned in Section 4.1.3.

Extreme chemical weathering is most effective in subtropical regions where high seasonal rainfall leads to repeated fluctuations in the water table (Section 2.3). The resultant variations of Eh and pH facilitate the breakdown of feldspar and ferromagnesian silicates and removal of alkalies to leave kaolin and a variety of oxides and hydroxides. A hard, nodular or earthy crust, resistant to weathering and erosion, remains after intensive leaching. Laterite is enriched in ferric iron oxides and hydroxides and is rusty red, brown or yellow in colour. It is very widespread on old weathering surfaces formed at low latitudes and is used frequently as a building material (Section 5.2) and locally as an iron ore (Fig. 4.8).

Bauxite, consisting mainly of aluminium hydroxides, is considerably less widespread and is developed mainly on granites, feldspathic sandstones and clays which have a low iron content. Tertiary bauxites formed by repeated solution, redeposition and resolution in the zone of weathering occur in Jamaica and near the coast of Surinam. Those of the type locality Les Baux, in France, are

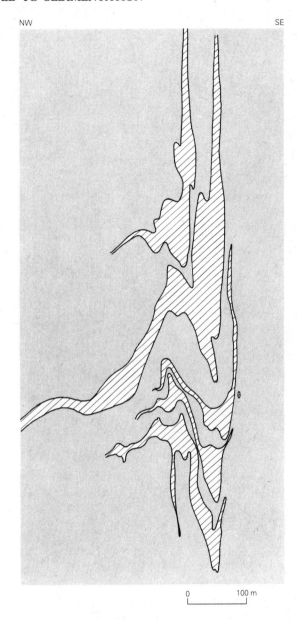

NW   SE

0   100 m

Figure 4.9 Deformation of ore bodies during metamorphism Illustrated by a sulphide Ag–Pb–Zn ore body at Broken Hill, New South Wales: the lenticular sulphide bodies are concordant with the banding of the metamorphic host rocks, details of which are omitted for clarity (based on Carruthers, D. S. and R. D. Pratten 1961. *Econ. Geol.* **56**).

thought to have resulted from the leaching action of hot volcanic waters. Although alu-

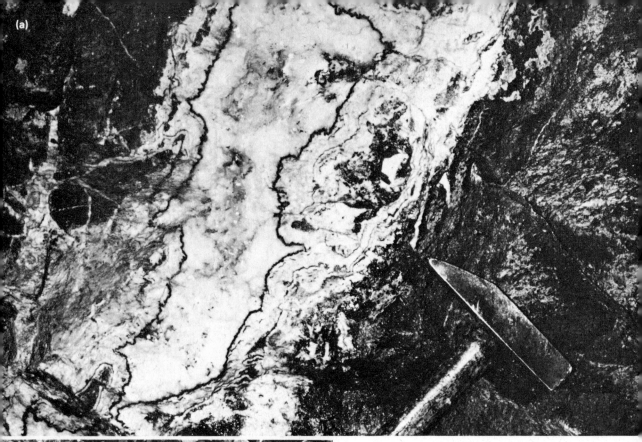

Figure 4.10 Hydrothermal veins (a) Crustified texture resulting from the deposition of successive layers on the sides of an opening fracture: from outside inwards, the principal minerals are scheelite ($CaWO_4$, an ore mineral of tungsten), carbonates, sphalerite (ZnS) and calcite (Carrock mine, Cumbria – photo by P. Garrard). (b) Quartz veins in echelon fractures opened by differential movement, illustrating the structural control of vein systems (photo by P. Garrard).

minium is very abundant (Table 4.2), it cannot as yet be economically extracted from the rock-forming silicates of which it is a component. Extraction from bauxite, the principal ore, is costly in terms of energy, for which reason aluminium smelters are commonly sited near sources of hydroelectric power. When demand outran supply in World War 2, production in the USA was trebled by the use of normally uneconomic low-grade ores.

Metals released from sulphide ores during weathering are carried by ground waters down through the zone of oxidation, leaving a depleted surface crust (or gossan) composed largely of red, yellow and brown iron oxides and hydroxides. At or near the water table changes in the chemical environment lead to the redeposition of Cu, Zn, Pb and Fe in a zone of **secondary enrichment** higher in grade than the unmodified ore body at depth. Oxides or hydroxides may be mixed with sulphides in this zone.

4.5 Metamorphism and deformation in relation to mineralisation

Many metamorphic terrains contain ore bodies and metamorphism has often been thought of as an important ore-forming process in its own right. Recent investigations suggest, however, that comparatively few deposits have been formed *ab initio* by metamorphism. The most important deposits so formed are the pyrometasomatic ores at the contacts of plutonic intrusions, which have been mentioned elsewhere (Section 4.3.3).

Other, more widely developed deposits represent pre-existing ore bodies which have been metamorphosed along with their host rocks. A well known example of this type is the large stratiform massive sulphide (Pb–Zn–Ag) deposit of Broken Hill, New South Wales (Fig. 4.9). The enclosing rocks are strongly deformed pyroxene granulites and amphibolites (derived from basic igneous rocks) and garnetiferous gneisses (derived from the andesitic volcanics or sedimentary rocks). The ore bodies, made largely of coarse-grained sphalerite and galena, form concordant lenticles which reach their maximum thicknesses of 50–100 m in the hinge zones of several tight folds. Structural and textural evidence suggests that the ore bodies were subjected to repeated deformation and metamorphism along with their host rocks and implies, therefore, that mineralisation preceded metamorphism; the original nature of the ores has been obscured by later modifications – they may have been of exhalative or sedimentary origin.

Sulphide or oxide ore bodies respond to high temperature and high stress by recrystallisation or development of new minerals, by coarsening of grain size and by distortion, folding or disruption of the ore body (Fig. 4.1). These changes inevitably modify the grade and three-dimensional form of an ore body in ways which are of critical importance to the miner. For example, the continuity of a stratiform ore body may be interrupted and unexpected contortions and variations of thickness may be encountered, as is illustrated for the silver–lead–zinc sulphide deposit of Broken Hill (Fig. 4.9).

4.6 Hydrothermal vein systems

The role of hot aqueous solutions as mineralising agents has been emphasised more than once. Such hydrothermal solutions may derive a variety of metals from magmatic sources, from sedimentary basins or from previously mineralised terrains. They are commonly saline and they range in temperature from ~560 °C to ~150 °C. Vein deposits are important sources of gold, both directly and via the formation of gold placers (Section 4.4.1), and of lead, zinc, silver, tungsten and bismuth (Table 4.9).

Deposition from solution commonly takes place in response to changes in temperature, pressure, pH or Eh encountered where migrating fluids pass from one host rock to another or where fluids from different sources meet. Replacement of the host rock plays a part in the formation of ores in many environments. Vein deposits are, naturally, most often formed high in the crust where open fractures can exist and tend to concentrate near faults.

The structural and textural features of hydrothermal veins commonly record the infilling of an opening fracture or a pre-existing void by successive outgrowths from the walls (Fig. 4.10). Each crystalline coating serves as a base for growth of further minerals, so that a layered structure (**crustification**) is built up,

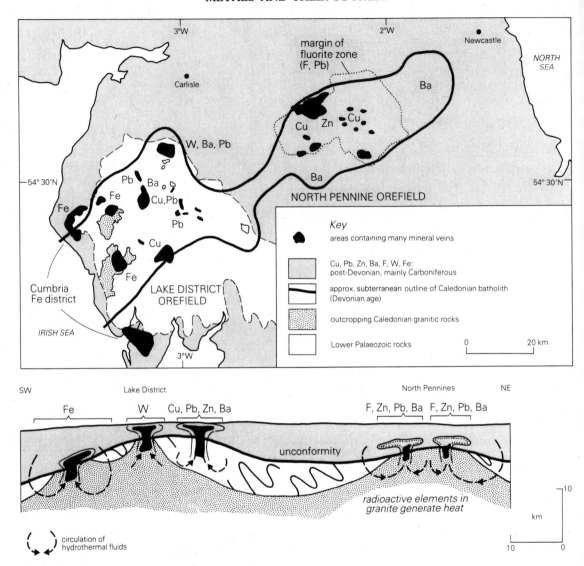

Figure 4.11 Hydrothermal orefields in northern England The location of vein deposits (above) in relation to the pre-Carboniferous basement granites of the Lake District and North Pennines: and an hypothetical section (below) showing conditions at the time of mineralisation – the Cumbrian iron ores consist mainly of haematite (modified from Moore, J. McM. 1982. *Metallization associated with acid magmatism.* New York: Wiley).

sometimes symmetrically arranged relative to the vein walls. **Comb structure** is produced by the parallel growth of quartz, calcite or other prismatic minerals which project as well formed (idiomorphic) crystals into the unfilled centre of the fracture or void. Alteration of the wall rock caused by the passage of fluids leads to replacement of minerals by densely packed growths of fine-grained chlorite, zoisite, sericite, calcite or clay minerals.

The character of the ores and the nature of the gangue minerals depend on the one hand on the genetic affinities of the hydrothermal fluids and, on the other, on the temperature of deposition. Examples of several genetic types have been mentioned in previous sections

Table 4.9 Hydrothermal vein deposits.

Vein assemblage	Common minerals	Geological setting	Examples	Principal metals
precious metals	gold tellurides, native gold, quartz, sulphides	volcanic provinces, especially early Precambrian greenstone belts and Tertiary circum-Pacific provinces T 600–100 °C	Cripple Creek, Colorado (Tertiary) Kirkland Lake, Ontario (Precambrian)	Au Ag
polymetallic sulphides	sulphides, quartz, carbonates, tourmaline	in subvolcanic settings and around granitic plutons T 500–100 °C	Western USA, SW England (Hercynian)	some or all of: Cu Zn Pb Ag Au Fe Mo (W Sn)
lead–zinc	sphalerite–galena, quartz, fluorite, baryte, calcite	in carbonate hosts not directly related to igneous centres T<200 °C	Mississippi-Valley-type ores	Pb Zn
cobalt type silver	sulphides, arsenides, native silver, bismuth quartz, calcite	rare, near basic igneous intrusions	Cobalt district Ontario (Precambrian)	Ag Co Ni Bi

and the features of these and other types are summarised in Table 4.9. Deposits formed at relatively high temperatures are more or less closely related to volcanic or plutonic centres which acted as heat sources; lower-temperature deposits, such as the Mississippi-Valley-type lead–zinc deposits, may be located in sedimentary basins (Section 4.4.3). A regional zoning with respect to the distribution of both ore and gangue minerals which relates to the temperature of deposition is developed around the Hercynian batholith of south-west England (Fig. 4.4) and around the Bingham, western USA, porphyry copper (Fig. 4.5). The North Pennine orefield of England, in which the veins were emplaced in Carboniferous strata at about the end of the Palaeozoic era, shows a remarkable concentric arrangement centred over a buried pre-Carboniferous basement granite rich in radioactive elements, which appears to have acted as a heat source controlling a convective hydrothermal circulation system (Fig. 4.11).

References

Cronan, D. S. 1980. *Underwater minerals*. London: Academic Press.

Derry, D. 1980. *World atlas of geology and mineral deposits*. London: Mining Journal Books.

Dixon, C. J. 1979. *Atlas of economic mineral deposits*. London: Chapman & Hall.

Evans, A. M. 1980. *An introduction to ore deposits*. Oxford: Blackwell Scientific.

Jensen, M. L. and A. M. Bateman 1979. *Economic mineral deposits*, 3rd edn. New York: Wiley.

Knill, J. L. (ed.) 1978. *Industrial geology*. Oxford: Oxford University Press.

Open University 1976. *The Earth's physical resources: Mineral deposits*, S266, Block 3. Milton Keynes: The Open University.

Open University 1976. *Earth science topics and methods*. Porphyry copper case study, S333. Milton Keynes: The Open University.

Read, H. H. 1970. *Rutley's elements of mineralogy*, 26th edn. London: George Allen & Unwin.

Rose, A. W., H. H. Hawkes and J. S. Webb 1979. *Geochemistry in mineral exploration*. London: Academic Press.

Thomas, L. J. 1973. *An introduction to mining*. Sydney: Hicks Smith.

Warren, K. 1973. *Mineral resources*. London: Penguin.

71

5 Non-metallic raw materials

5.1 Introduction

The mineral and rock materials dealt with in this chapter are extremely diverse, both in geological affinities and in the uses to which they are put; they range from common substances like clay, extracted in enormous volumes, to such rarities as diamond, mined only at a handful of sites. The grouping of products adopted here is simply one of convenience. The materials used in the construction industry are dealt with first, since they are the most important resources in terms of bulk. The raw materials of the ceramics industry follow naturally, because they include many of the same substances. Organic chemicals and synthetics derived from hydrocarbons are dealt with under a single heading, despite the varied nature of the end products, and the deposits of evaporitic environments are similarly considered as a group. Lastly, a miscellany of useful and decorative materials are mentioned, with no pretensions to comprehensiveness.

It is important to realise the scale on which the principal non-metallic geological materials are extracted. A large office block requires about 2000 tonnes of aggregate for concrete and a kilometre of new road at least 10 000 tonnes. About 100 million tonnes of sand and gravel were extracted annually in Britain in the early 1970s, roughly equalling the annual coal production of the same period (Fig. 5.2). Bulk construction materials are generally extracted from open pits and quarries which, because of the high cost of transport, tend to be located as near as possible to construction sites. The need to maintain supplies and the environmental problems posed by the rehabilitation of old pits and quarries has stimulated the development of offshore extraction sites in estuaries and other sheltered waters.

5.2 Materials for construction
(Table 5.1)

5.2.1 Stone and slate

Dimension stone was for centuries the principal load-bearing material of buildings, bridges, harbour works and so on, a function now largely taken over by concrete and steel. The common building stones are granites and massive sandstones and limestones which can be quarried in sizeable rectangular blocks free from internal fractures, without yielding an undue proportion of waste fragments. High compressive and shear strengths are required for load-bearing structures (Table 6.4). A wider variety of porphyritic igneous rocks, marbles, tuffs (such as the Borrowdale Volcanics of Cumbria), fossiliferous limestones and travertine is used as decorative stone for facings, pavings and interior walls. Slate, characterised by a closely spaced cleavage developed by crustal stresses, which facilitates the separation of thin layers, is a traditional roofing material, the principal British source being in the Cambrian slates of North Wales. The bulk and weight of dimension stone required for major building works demand ease of transport from quarry to construction site. In Britain, 18th- and 19th-

Table 5.1 Non-metallic construction materials.

Product	Sources	Desirable properties
dimension stone	limestone, sandstone, granite, other igneous rocks. Ornamental stone includes limestone, marble, tufa, granite, syenite etc.	regular, widely spaced partings (bedding, joints), high compressive strength, resistance to weathering, especially in industrial regions
slate	strongly cleaved fine-grained metamorphic rocks, usually of pelitic composition, locally pyroclastic	regular, closely spaced cleavage, resistance to weathering
roadstone	crushed basalt, dolerite, fine granite, greywacke, quartzite, hornfels, flint etc., industrial waste in combination with bitumen	resistance to abrasion (massive, fine to medium grain size), low porosity, binds well with bitumen, non-slip surface, does not acquire polish
aggregate (for concrete and as fill for road and building foundations, dams)	sand and gravel (fluvial, glacial, marine), crushed rock as for roadstone, industrial waste	appropriate range of particle sizes, low contents of impurities, especially sulphides, organic matter, coal, micaceous rocks, opal, chalcedony
bricks, tiles	clay, marine, alluvial, glacial or in deep weathering zones: raw materials fired at high temperatures	no excess water, low iron, sulphides, sulphates $CaCo_3 > 5\%$ minimises shrinkage, carbonaceous matter ($\geqslant 5\%$) assists firing
cement	limestone, argillaceous limestone, often mixed with clay: limestone converted to lime by calcining in kiln, product ground to powder	constant composition, correct ratios CaO, Al_2O_3, SiO_2, Fe_2O_3, low S, MgO, P, alkalies
glass	quartz sand, quartzite	absence of impurities, low iron
plaster, plasterboard	gypsum, anhydrite from evaporites (see Section 6.4)	—
insulating materials	fibrous and flaky metamorphic minerals, asbestos, mica, vermiculite: diatomite	not injurious to health (see Section 7.5)
bitumen	residue from distillation of oil: natural residues of oil seepages	appropriate 'melting' temperature for conditions of use, e.g. in road making

century quarries in the Channel Islands and Aberdeen (granite) and southern England (Portland Limestone) were sited on the coast, others (Shap and Dartmoor granites) on main railway lines.

The most durable building stones are granites and similar plutonic rocks with massive texture, low porosity and stable minerals (Fig. 5.1). Sandstones, especially those with calcareous cement, are subject to the effects of permeation by water, and limestones to solution and reaction on a larger scale. Despite these disadvantages, limestones have provided some of the most beautiful building materials of southern Britain where the creamy Middle Jurassic limestones characte-

rise many buildings in Bath, Oxford and the Cotswold villages, and Upper Jurassic Portland Stone is prominent in St Paul's cathedral and other London churches built by Wren.

The conservation of stonework in historic buildings and monuments presents problems which depend on both the properties of the rock and climatic factors. Wind abrasion is a principal cause of damage for the sphinxes and other monuments of ancient Egypt. In north-west Europe, decay is due largely to the activity of water in pore spaces and incipient fractures. Mechanical disintegration due to freeze and thaw is increased by chemical reactions when waters pick up dissolved CO_2, H_2S and SO_2 in regions where the

Figure 5.1 Durability of stone Weathering in urban environments, illustrated by contrasting examples from St Paul's Cathedral, London. (a) A bust of St Andrew, formerly located on a pediment at the west front: the limestone (Portland Stone) is severely corroded. (b) A bollard close to the west front: the Shap granite is almost unaffected. (Photos by M. Gray).

atmosphere is polluted by industrial processes (Section 7.6). Acid rain water attacks carbonates in particular and sulphates formed in the presence of SO_2 may give rise to ugly efflorescences of gypsum ($CaSO_4.2H_2O$). Damage by such processes and blackening due to the fixing of soot encrustations after the Industrial Revolution have been slowed down in Britain by legislation to control emissions from chimneys.

5.2.2 Aggregate

Aggregate, which bulks out concrete and provides foundations for roads, dams and harbour works, can be supplied from many sources both natural and artificial. Sand and gravel, the most common components, are taken wherever possible from unconsolidated glacial, alluvial or marine sediments which can be excavated cheaply by mechanical scoops and dredges.

Although sand and gravel are widely available, many deposits are too small or too variable to repay working. The assessment of resources involves the detailed mapping of glacial, alluvial and coastal deposits (often lumped together as 'drift' on small-scale geological maps). Figure 5.2a illustrates the preliminary identification of resources in an area where new industrial development is anticipated. Up to 13% of the sand and gravel used in Britain in the early 1970s was obtained from offshore sites, mainly by the use of suction dredges.

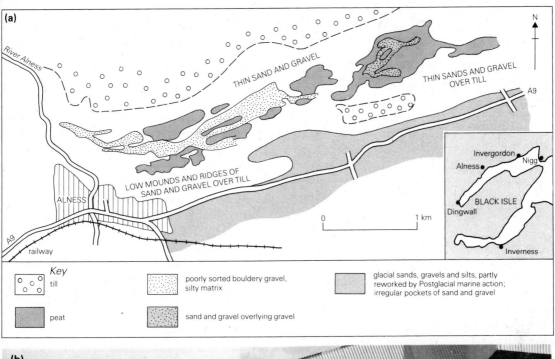

(a)

THIN SAND AND GRAVEL

THIN SANDS AND GRAVEL OVER TILL

River Alness

LOW MOUNDS AND RIDGES OF SAND AND GRAVEL OVER TILL

ALNESS

A9

railway

N

A9

Invergordon

Alness

Nigg

BLACK ISLE

Dingwall

Inverness

0   1 km

Key

till

peat

poorly sorted bouldery gravel, silty matrix

sand and gravel overlying gravel

glacial sands, gravels and silts, partly reworked by Postglacial marine action; irregular pockets of sand and gravel

Figure 5.2   Sand and gravel   (a) The Alness region in north-east Scotland is within reach of areas where industrial developments were expected in the 1970s (Nigg is an oil rig construction site). The sketch map shows the extent of ridges of glacial sand and gravel suitable for exploitation; these ridges are deposits of streams flowing beneath or alongside an ice sheet. Proximity to several roads and streams provides for easy access and for washing of sand and gravel during extraction (based on Harris, A. L. and J. D. Peacock 1969. *IGS Report* 69/9). (b) The scale on which sand is extracted is illustrated by this photo of equipment for washing and grading at a pit in Northern Ireland (Crown Copyright reserved).

Aggregate derived from igneous, metamorphic and lithified sedimentary rocks is normally crushed and graded at the site of quarrying. Laterite is an important resource in the Tropics, and spoil from mines, slag from smelters and a variety of other waste industrial materials can (provided they are not chemically active) be used where they are available. The considerations that determine the choice of material for a given purpose include: (i) costs, depending largely on scale and accessibility of source material and ease of extraction; (ii) physical properties, especially the proportion of coarser and finer particles (too high a proportion of clay hinders bonding with cement or bitumen) and the shapes and surface characteristic of the larger fragments (flaky or elongated fragments weaken the structure, over-smooth surfaces do not bond well and materials that take a polish promote skidding when used as road metal); and (iii) presence or absence of reactive impurities (certain clays tend to swell when wet, sulphides oxidise with release of H_2S, chalcedony and opal convert to gels in the presence of alkalies). Where materials with the required properties are scarce, it may be necessary to remove undesirable components such as excess clay by washing or more complex forms of treatment.

5.2.3 Cement

Cement is an artificial powder which sets rapidly to a rock-like hardness after being mixed with water. When combined with sand in mortar, it is used to seal brickwork or stonework. Bulked out with large volumes of aggregate, it provides the setting agent in concrete.

The principal component of cement is calcium carbonate from which a proportion of the CO_2 has been expelled by heating. A rather primitive cement can be obtained by roasting and grinding many types of limestone and argillaceous limestone. Portland cement, a 19th-century invention, is made by heating limestone with clay to temperatures high enough for calcium silicates to be formed by reaction between clay minerals and calcium carbonate. The resulting stony clinker is powdered and must, of course, be kept dry until it is used.

The principal sources of cement in Britain are Chalk (mixed with Mesozoic or Tertiary clays) in the south-east and Carboniferous Limestone (mixed with Carboniferous shales) elsewhere. The principal components are lime, silica and alumina and the ratios of these components can vary only within certain limits. High alumina cement has in certain instances been associated with mechanical failure of concrete. Impurities which affect quality include magnesium, a component of dolomite which converts to periclase (MgO) at high temperatures and subsequently hydrates to brucite ($Mg(OH)_2$) with an increase in volume. Sulphides and sulphates convert to alkali sulphides with change of volume, while iron, manganese, phosphorus and fluorine are prone to reactions which weaken the end product.

5.2.4 Bricks and tiles

Bricks have been made from baked clay for more than three thousand years and commercial production in Britain now runs at several thousand million bricks per year. A satisfactory brick clay has a fairly low water content (17.5–19%) which limits shrinkage during drying. Small proportions of calcium carbonate and silt increase strength and reduce shrinkage, small amounts of iron give a pleasing colour, and organic matter, which releases energy when burned, reduces fuel costs during firing. These requirements are best met in Britain by the Jurassic Oxford Clay on which many brickworks are situated. Other Mesozoic and Tertiary clays are also used in Britain and bricks of a kind can be made from a variety of materials. Laterite (Latin *later*, a brick) supplies durable building blocks in subtropical regions (for origin see Section 4.4.6) and was used in the construction of many temples and palaces in

Cambodia and Thailand. Refractory bricks used for furnaces where very high temperatures are reached are made from kaolinite clay.

5.2.5 Glass

Natural and man-made glass are supercooled liquids in which extremely high viscosity has prevented the ordering of ions into crystal lattices. Glass is a transparent or translucent isotropic substance which breaks with a conchoidal fracture. The glass of commerce is generally selected for its transparency, homogeneity and lack of colour. It is essentially a silica glass, produced by the fusion of quartz (SiO_2) in the presence of additives such as calcium or sodium carbonate or sodium sulphate. Impurities tend to impart a colour ('bottle green' is largely due to the presence of iron) and to reduce the homogeneity of the product. Hence, glass sands are sought among well differentiated, well sorted sediments, mainly of shelf facies. Quartz sand, sandstone with silica cement and metamorphic quartzite are the principal source rocks.

5.2.6 Plaster, plasterboard and insulating materials

The plaster and plasterboard used to impart a smooth finish to walls and ceilings are derived mainly from the mineral gypsum (hydrated calcium sulphate, $CaSO_4.2H_2O$). When gypsum is heated at 110–120 °C, a part of the combined water is lost and the resulting substance yields a white powder, 'plaster of Paris', which sets hard after being mixed with water. The plasterboards now more commonly used than the simple powder, are generally impregnated with plaster mixed with kaolin. Gypsum is one of the evaporite minerals dealt with in Section 5.5.

Insulating or fire-resistant materials have for many years been obtained mainly from the minerals known collectively as asbestos. The distinctive property of these minerals is a fibrous habit which yields needle- and thread-like crystals capable of being spun, or compacted into felts and boards. The principal mineral species yielding asbestos belong to the serpentine group (chrysotile, formed along with other serpentines where ultrabasic rocks have undergone low-grade metamorphism) and to the amphibole group (asbestiform tremolite, found in low-grade metamorphic rocks, and the sodic crocidolite which forms veins in banded iron formations (Section 4.4.4)). Evidence that asbestos fibre tends to enter the tissues of the human lung has cast a shadow over the future use of this resource (Section 7.5). Other natural insulating materials include diatomite (a lacustrine deposit composed of the siliceous tests of microscopic plants), vermiculite (a chlorite-like mineral from low-grade metamorphosed ultrabasic rocks), and powdered mica.

5.2.7 Bitumen and asphalt

A variety of dark, viscous hydrocarbons occur naturally in pits and 'lakes' from which more volatile oils escaping from the Earth have evaporated. Asphalt obtained from such sources – for example in Trinidad – has long been used as a waterproofing material and to bind rock chips in a 'metalled' or 'tar macadam' road surface. Dark tarry substances produced during the distillation of crude oil (Section 3.5.2) are used for similar purposes.

5.3 Ceramics, refractories and fillers

Pots and other domestic utensils have been made from plastic clay, moulded and dried either by exposure to sunlight or by heating in a kiln, since very early times; indeed, pottery fragments are among the commonest artefacts of prehistoric sites and they provide some of the best guides to age and cultural affinity. Earthenware of a kind can be made from almost any clay and local sources are used for simple culinary utensils, drainpipes, tiles, flowerpots and similar objects. The colour varies from white to red. Earthenware remains porous after firing but is commonly sealed by a glaze applied before or part way through the process of firing. Terracotta (unglazed), delft and faience (glazed) are well known varieties of earthenware pottery.

Whiteware, which includes most fine domestic china as well as insulators and laboratory ware, is derived from clay characterised by a high proportion of the clay mineral kaolinite, $Al_4Si_4O_{10}(OH)_8$. In the making of stoneware and porcelain, the clay, diluted with water, is mixed with natural or added 'fluxes' such as powdered flint, quartz, and feldspar and after shaping the mixture is partially vitrified by firing at temperatures of about 1200 °C. Chinaware or bone china contains calcium phosphate derived from calcined bone. All these forms are almost white, dense and impermeable, bone china being slightly translucent. Colour, finish and lustre depend on the kind of glaze added during the firing process.

The purest kaolinitic china clay is formed not by sedimentary processes but by hydrothermal or pneumatolytic alteration of granite. In Britain china clay of very high quality (an important export for two centuries) occurs as pipe-like or irregular masses in the Hercynian granites of south-west England (Fig. 4.4). Within these areas, the passage of hot, chemically active fluids has led to the kaolinisation of feldspar and the breakdown of granite into a friable material containing remnants of quartz. Kaolin is washed out from pits by high-pressure jets of water and separated from the quartz in settling tanks. St Austell (Cornwall) is now the main centre of production. Less pure ball clays derived from china clay by erosion and redeposition are found in early Tertiary lake beds at Bovey Tracey and Petrockstow in Devon (Fig. 3.16).

Fillers used to add bulk or weight to paper, rubber and synthetic products are derived mainly from inert clay minerals and from baryte ($BaSO_4$), used also as a component of oilfield drilling muds. Baryte is a common gangue mineral in hydrothermal and exhalative sulphide deposits and it occasionally forms larger concentrations.

Materials that retain their shape and chemical identity without marked expansion at high temperatures are required for lining kilns and furnaces and for many purposes in the electrical and chemical industries. These refractories and the technical ceramics which

Table 5.2 Refractories and technical ceramics.

Raw material	Source	Product and use
fireclay	in Britain, seat earths of Coal Measures (Table 3.5)	refractory bricks for domestic fireplaces, furnaces
quartz sandstone, quartzites	in Britain, mainly from Upper Carboniferous ganisters	silica bricks for furnaces
dolomite rock	in Britain, mainly Magnesian Limestone (Permian) Carboniferous Limestone	basic refractories stable in presence of slags, for iron and steel furnaces
magnesite ($MgCO_3$)	low-grade alteration of ultrabasic rocks (e.g. Greece, Yugoslavia); replacements in limestone (e.g. Urals)	basic refractories stable in presence of slags, for iron and steel furnaces
serpentine, olivine rock	ultrabasic igneous rocks	
mullite	aluminium silicates (sillimanite, kyanite, andalusite) from metamorphosed pelites; mullite ($3Al_2O_3 . 2SiO_2$) formed by heating to 1500 °C	sparking plugs, other electrical equipment
alumina, Al_2O_3	bauxite and related rocks (Section 4.4.6), treated with caustic soda	technical ceramics
talc, steatite	metamorphosed magnesian limestone	insulators in radio industry
oxides of Be, Zr, Th, Mg	beryl, zircon, monazite etc. from pegmatites, placers	technical ceramics

require a range of specialised properties are derived from both natural and synthetic sources. The more important natural sources are listed in Table 5.2.

5.4 Organic chemicals and synthetics

Until little more than fifty years ago, all but a minute proportion of the oil, gas and coal extracted from the Earth was used as a fuel. Today, a substantial fraction of world production (mostly of hydrocarbons) goes to provide feedstock for the production of organic synthetics (Table 5.3). The petrochemical industry produces relatively cheap alternatives to many natural organic and inorganic materials. The technology involved in the manufacture of these substances is complex and only the principal processes employed are mentioned here. The lower unsaturated olefins such as ethylene, C_2H_4 (Section 3.5.3), provide the starting points from which many synthetic molecules are made. These olefins are obtained largely from the paraffins in crude oil by high-temperature

Table 5.3 Derivatives of hydrocarbons and coal (other than fuels).

Refinery products
extracted from crude oils and their derivatives:
    lubricating oils
    bitumens (see Table 5.1)
    waxes (used mainly for waterproofing)
    detergents (refinery products mixed with other
      chemicals)

Breakdown products
ammonia and ammonium salts made from hydrogen in
    combination with atmospheric nitrogen, starting point
    for synthesis of nitrogenous fertilisers (Table 7.2)

carbon made by high-temperature dissociation, used to
    strengthen synthetic rubber and for carbon fibre

Polymers of hydrocarbon derivatives
    plastics, e.g. polyethylene, PVC, polystyrene
    silicones
    synthetic fibres, e.g. nylon, terylene, acrylic fibre
    dyes and paints
    pharmaceuticals
    insecticides
    aerosol propellants
    explosives

cracking procedures which rupture long carbon chains and remove excess hydrogen; they can be subsequently combined with molecules containing oxygen or chlorine in addition to carbon and hydrogen. Once formed, these varied compounds provide building blocks for the synthesis of high polymers in which thousands or tens of thousands of atoms are linked. **Polymerisation** gives analogues of molecules in wood, silk, cotton, rubber and other natural substances for which the synthetic materials can deputise. A selective list of products is given in Table 5.3.

5.5 Derivatives of evaporites

Minerals that are soluble in water are deposited only in unusual geological conditions where natural waters have been concentrated to saturation point by evaporation (Table 2.1). The evaporites, the chemical sediments of arid climatic zones, are source materials for a wide range of chemicals, fertilisers and other commodities (Table 5.4; see also Table 7.2). Marine evaporites deposited from highly concentrated sea water accumulate on the floors of gulfs and bays partially cut off from the open sea and in sabkha zones fringing the sea where unconsolidated sediments permeated by brines are enriched in salts as water is lost by evaporation. In both environments, the deposits consist principally of calcium sulphates (gypsum, $CaSO_4.2H_2O$, or anhydrite, $CaSO_4$) and rock salt (NaCl), with a small but valuable proportion of potassium and magnesium salts (bittern salts) and with significant concentrations of bromine and iodine. Evaporites of non-marine salt lakes formed in areas of interior drainage where evaporation exceeds inflow, yield mineral suites dependent on the composition of the rocks exposed in the drainage basin (Table 5.4). Modern brines and evaporites in arid climatic zones are processed directly for salt and other products. Ancient evaporites have a different distribution, because continental drift has

Table 5.4 Evaporite minerals and their uses.

Mineral	Occurrence	Uses
Sulphates		
gypsum $CaSO_4.2H_2O$ anhydrite $CaSO_4$	relatively low solubility ensures early deposition, common in marine and sabkha environments, may occur without other evaporite minerals	gypsum is raw material of plaster and plasterboards (Section 5.2); other uses of both sulphates include production of sulphuric acid, insecticides and fertilisers
epsomite $MgSO_4.7H_2O$	late-stage deposit of marine evaporates	pharmaceuticals
mirabilite $Na_2SO_4.10H_2O$	non-marine soda lakes	manufacture of paper, textiles, dyes, explosives, fertilisers
Native sulphur S	derived from gypsum – anhydrite by bacterial action in salt domes	sulphuric acid, fertilisers
Halides		
halite, rock salt NaCl	follows gypsum but precedes potash salts in marine evaporite sequence	essential in nutrition, also preservative, source of caustic soda, soap, dyes, insecticides, drugs, etc.
sylvite, KCl	late replacement mineral in potash deposits	sylvinite (sylvite with halite) is an important potash fertiliser
carnallite $KCl.MgCl_2.6H_2O$	late (bittern) stage in marine evaporites	fertiliser
iodates of Ca, K, Cr	byproduct of Chilean nitrates	antiseptic, source of iodides and compounds used in photography
Carbonates natron $Na_2CO_3.10H_2O$	non-marine soda lakes	soap, glass, caustic soda
Borax $Na_2B_4O_7.10H_2O$	non-marine evaporites where igneous rocks or hot springs supply boron to catchment area	used, with other boron minerals in glazes, enamel, glass, paper and leather manufacture
Nitrates nitre KNO_3 soda nitre $NaNO_3$ complex salts of K, Na, Mg	non-marine evaporites, mainly from Chile	fertilisers, explosives, nitric acid

caused the continents to migrate northwards through the arid climatic zones. The principal economic deposits are those in the Permo-Triassic succession in Europe where Zechstein (late Permian) deposits are of particular importance (Fig. 5.3) and in the mid-Tertiary of both Europe and North America.

Ordinary sea water (Table 2.1) contains only about 3.5% of dissolved matter, most of which is derived ultimately from the weathering of exposed rocks on land. Precipitation of even the least soluble minerals does not begin until the water volume has been reduced by half, while the most soluble bittern salts of potassium and magnesium are deposited only when evaporation is almost complete and are readily dissolved again

during diagenesis or weathering. Marine evaporite sequences therefore show a characteristic succession from gypsum or anhydrite to thick rock salt, and finally to the valuable bittern salts which are rarely preserved. Interruption of the process of concentration by the influx of ordinary sea water may lead to the repetition of several cycles of evaporation and deposition, as in the important deposits of Stassfurt in Germany (Fig. 5.3). Other major sources of potash salts lie in Permian evaporites of New Mexico and in early Tertiary basins of the Rhine graben and Ebro basin in Spain. Rock salt (halite, NaCl), an essential mineral for human and animal nutrition and an article of commerce since prehistoric times, is more widely distributed;

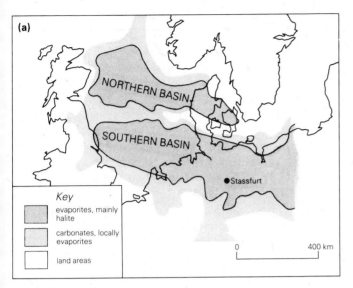

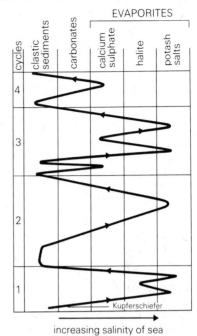

(b) Zechstein standard succession

increasing salinity of sea

Figure 5.3 Geological setting of Permian potash deposits (a) The saline Zechstein sea deposited evaporites in two basins extending across the North Sea and adjacent parts of Europe: evaporites of the southern basin, which locally reach more than 100 m in thickness, show up to four cycles of deposition. (b) The standard succession is based on the Stassfurt area: potash deposits in northern England probably correspond to Zechstein 2 – note the stratigraphical position of the Kupferschiefer, mentioned in connection with its copper deposits in Chapter 4.

some of the principal sources are in the Triassic of Europe.

Non-marine evaporites are less constant in composition, the species and relative importance of the minerals present depending on the chemistry of rocks in the drainage basin and on contributions from volcanic exhalations and other sources (Table 5.4). Soda lakes characterised by sodium carbonates occur in the arid sector of the East African rift valley where alkaline and mafic volcanicity is widespread. Borax and related minerals characterise the salt lakes of Utah and other mid-western states of the USA. Finally, the unique nitrate deposits of the Atacama desert in Chile have been major sources of sodium and potassium nitrates and of iodides, together with borates, sulphates and rock salt. These deposits, distributed over a 700 km tract west of the Andes, appear to owe their origin to extreme aridity persisting since

mid-Tertiary times and to a paucity of nitrate-using plants and bacteria. Contributions from Andean volcanicity account for the presence of some minerals. The saline constituents and nitrates may, according to a recent hypothesis, derive ultimately from Pacific spray rich in organic nutrients that yield ammonia, deposited as dry fall-out or condensate from fog.

5.6 Phosphorites

Phosphorus is essential to life processes, and phosphatic fertilisers are among the principal aids to increased crop production (Table 7.2). Over 80% of such fertilisers are derived from phosphatic sedimentary rocks, the principal exporters at present being Morocco, Tunisia, the USA, the USSR and Togo. Uranium, rare-earth elements, vanadium, fluorine and even gold show anomalous concentrations in some phosphorites and may be byproducts of the extraction process.

81

The majority of phosphorites are marine deposits, commonly associated with organic matter. They form pale earthy, often nodular rocks in which the mineral apatite $(Ca_5(OH,F,Cl)(PO_4)_3)$ carries much of the phosphate. Phosphorites (many apparently of mid-Tertiary age) occur *in situ* on the modern ocean floors off the coasts of California, northwestern and southern Africa, Australia, New Zealand and western South America. They appear to be located near the sites of nutrient-rich upwellings from deeper levels. Somewhat older deposits, for example the late Cretaceous – early Tertiary strata of Morocco, appear to have been formed in shallow embayments where organic materials were reworked at low oxygen concentrations. The remarkably widespread late Precambrian–Cambrian phosphorites, many of which have only recently been identified, show that contributions from advanced animals are not required for the accumulation of phosphates. Phosphorites of this age in Vietnam, China, the USSR, India, Brazil, north-central Africa and Australia provide a resource of great potential value for developing countries.

5.7 Precious and decorative stones

Minerals and rocks that are valued more for aesthetic than for practical reasons range from the gemstones of state regalia to the ochres and umbers from which pigments were formerly derived. The role of such natural materials in the artistic development of successive civilisations must have been enormous and, although many of their functions have now been taken over by synthetic products, they merit brief consideration.

Gemstones derived from single crystals of a wide range of mineral species owe their attractions to their colour, transparency and capacity to withstand cutting and polishing. Diamond, ruby and sapphire (Table 5.5) have tight lattice structures, no well developed cleavage, high refractive indices and brilliant lustre which, in diamond, is enhanced by traditional methods of cutting that ensure the internal reflection of light (Fig. 5.4) and by strong dispersion of red and blue light. Exceptional hardness (diamond 10, corundum 9 on Mohs' scale) is the basis of the industrial uses of the minerals as abrasives and in cutting tools. Garnet, zircon, beryl and topaz (Table 5.5) are examples of more common minerals that only occasionally attain gem quality. Strong or distinctive colours in normally pale minerals such as quartz may be due to the incorporation of characteristic trace elements (beryl falls in this category – the colour of emerald is generally associated with Cr or V, that of sapphire with FeTi) or to peculiarities of the lattic structure; in zircon and certain other minerals, colour may be enhanced by heating.

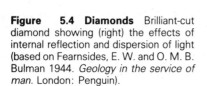

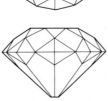

Figure 5.4 Diamonds Brilliant-cut diamond showing (right) the effects of internal reflection and dispersion of light (based on Fearnsides, E. W. and O. M. B. Bulman 1944. *Geology in the service of man.* London: Penguin).

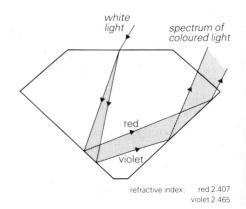

white light

spectrum of coloured light

red

violet

refractive index: red 2.407
violet 2.465

Table 5.5 Precious and semi-precious stones.

Name	Composition	Colour, other relevant properties	Occurrence	Principal localities
diamond	high pressure form of carbon	usually colourless, very high refractive index, density, hardness, dispersion	kimberlite, weathered kimberlite (blue ground), alluvial and beach placers	S. Africa, Namibia, W. Africa, Siberia
ruby sapphire	corundum, Al_2O_3	red blue, yellow	alkali basalts, silica-poor metamorphic rocks, placers	Cambodia, Thailand, Burma, Sri Lanka, E. Australia
emerald aquamarine	beryl $BeAl_2Si_6O_{18}$	green pale blue	pegmatites, granites	Colombia, Brazil, S. Africa
topaz	$Al_2SiO_4(OH, F)_2$	yellow, pink, blue	pegmatites, granites, alluvial placers	Urals, Brazil, Zimbabwe
zircon	$ZrSiO_4$	pale colours, intensified by heating	usually alluvial placers	Sri Lanka, Burma
garnet		red, green	metamorphic schists, skarns, alluvial placers	S. Africa
rock crystal amethyst cairngorm	quartz SiO_2	colourless, transparent purple smoky yellow	granites, pegmatites, hydrothermal veins	Brazil, Uruguay Scotland
jade	sodic pyroxene (jadeite), sodic amphibole (nephrite)	massive, felted fibres, green, grey	high-pressure metamorphic rocks	China, Burma, Guatemala, California, New Zealand
opal	hydrated SiO_2	play of colours, blue, green, red, amorphous	low-temperature replacements and void fillings	Queensland, Nevada, Mexico
agate	chalcedony (cryptocrystalline quartz plus opal)	concentric colour banding in grey, brown	veins and amygdale fillings in sediments and volcanics	many areas
Blue John	fluorite, CaF_2	massive, marbled blue, grey, purple	hydrothermal void fillings and replacements	Derbyshire (England)
turquoise	hydrated CuAl phosphate	peacock blue	oxidation of copper ores	India, Iran, Arizona
lapis lazuli	sulphur-bearing feldspathoid	massive, intense blue, marbled	contact-metamorphosed limestone	Iran, Afghanistan, Lake Baikal
alabaster	gypsum $CaSO_4.2H_2O$ calcite, $CaCO_3$	fine-grained, white or grey	evaporites, stalagmitic dripstone	Mediterranean region used in antiquity, probably from Egypt
amber	fossil resin	golden	estuarine Palaeogene sediments	Baltic region

Igneous rocks are primary sources of many gemstones. The rubies and sapphires of Cambodia and Thailand are contained in alkali–basalt necks. Peridot (olivine of gem quality, the chrysolite mentioned by Othello) is derived from phenocrysts in basic rocks. The primary source of diamond is kimberlite – an unusual ultrabasic igneous rock which forms dykes or cylindrical pipes apparently emplaced by a mechanism involving explosive gas blasting (for occurrence, see Section 4.3.1). The composition of kimberlite, and the fact that many intrusions carry nodules of eclogite or peridotite derived from the mantle, show that the parent magma originated at considerable depths. The experimentally determined stability field of diamond suggests that it is stable only at depths below about 150 km and, hence, it is supposed that diamondiferous kimberlite magmas have migrated from these depths carrying with them xenoliths and xenocrysts of mantle material which resisted partial melting. In addition to diamonds, the distinctive xenocryst suite includes magnesian garnet (pyrope) which acts as a marker in soils or stream sediments derived from weathered kimberlite.

Other gemstones originate in pegmatites of pneumatolytic origin (cf. Section 4.3.3) or in metamorphic rocks where concentrations of volatiles promoted the growth of large crystals. Such deposits tend to be small and patchy and are worked from open pits — indeed kimberlite is almost the only gem source valuable enough to justify the sinking of deep mines. The stability and high density of many gemstones ensure that they persist unchanged when the source rock is weathered (blue ground, the weathering product of kimberlite, is usually more productive than the parent rock) and, particularly, are segregated along with other heavy minerals during transport by water. Gem placers in southern Africa are major sources of diamonds along the Orange River and in old beach deposits of Namibia (Fig. 4.3); rubies, sapphires, zircon and other minerals characterise placers in Sri Lanka.

Among less valuable though still-decorative stones, used either in jewelry or in the manufacture of ornaments and mosaics, are massive mineral aggregates such as jade, lapis lazuli and alabaster, the cryptocrystalline material opal and agate, and a few, such as amber, of organic origin. The occurrence of these semi-precious stones is noted briefly in Table 5.5. Some of the same materials have found a use in mosaics and sculpture, together with certain rock types used as dimension stone (Section 5.2) and a few soft and easily worked rocks such as serpentine and steatite (soapstone or massive talc). Marble (metamorphosed limestone) however, was the principal statuary material used in the classic Greek and Roman periods and in the Italian Renaissance. Mesozoic limestones are widespread in southern Europe. Attica and the Cyclades were probably the main sources of marble for the Greeks; the Triassic Carrara Marble of the Apennines in Italy — an even-textured pure white rock subjected to low-grade Alpine metamorphism — has been quarried for over two thousand years.

A final group of decorative materials, more of historic than of present importance, is that of the pigments derived from mineral sources. Coloration is supplied by iron in various states of oxidation (ochres formed by oxidation of sulphides in deposits such as those of the Troodos range, Cyprus yield yellow, brown and reddish colours), by copper (mainly greens and blues from oxides and hydrated compounds), cobalt (blue) and lapis lazuli (ultramarine). The abundant base metal deposits available to the early Middle Eastern and Mediterranean civilisations made these natural sources of pigment readily available for frescoes and paintings.

References

Bates, R. E. 1965. *Geology of the industrial rocks and minerals.* New York: Dover.

Beaver, S. H. 1968. *The geology of sand and gravel.* London: Sand and Gravel Association.

Kirsch, H. 1968. *Applied mineralogy.* London: Chapman & Hall and Science Paperbacks.

Knill, J. L. (ed.) 1978. *Industrial geology.* Oxford: Oxford University Press.

McLean, A. C. and C. D. Gribble 1979. *Geology for civil engineers.* London: George Allen & Unwin.

Open University 1974. *Construction and other bulk materials.* S266, Block 4. Milton Keynes: The Open University.

Read, H. H. 1970. *Rutley's elements of mineralogy,* 26th edn. London: George Allen & Unwin.

6 Geological aspects of construction work

6.1 Stability of surface regimes

In many countries, the natural land surface has been extensively modified to suit the needs of human communities. Deforestation and farming have brought changes of vegetation, soil and drainage since prehistoric times and, more recently, civil engineering and mining works have modified coastlines, river systems and landforms on an even larger scale. All such changes involve the imposition of a partly artificial regime on existing geological and ecological systems. They tend to provoke reactions which may cause permanent and widespread damage if not foreseen and allowed for; among the best known of such reactions are the landslides that sometimes follow the construction of road or railway cuttings (Fig. 1.5) and the subsidence common in coal mining areas (Fig. 6.1).

Geological aspects of such disturbances of the natural equilibrium fall largely within the province of the engineering geologist. Parts of this large province are dealt with in Chapters

Figure 6.1 Road damage due to subsidence (Barnaby)

2, 5 and 7. Here, we shall deal with two main topics: (i) factors affecting the stability of natural and man-made structures, and (ii) consequences of major natural or artificial changes in landforms and land use. Discussion of both topics calls for understanding of the manner in which geological materials respond to the imposition of man-made structures – a matter involving the engineering properties of rocks and soils – and of the means of controlling potentially destructive developments in unstable situations. It should be remembered that not all geological hazards are man-made – the Earth's inherent instability is expressed by many natural changes which also threaten human safety.

6.2 Effects of instability

6.2.1 Mass movements under gravity

Gravitational forces are potent factors affecting stability of the Earth's surface in that water, soil and rock tend, when unsupported, to move down slope. The slow but inexorable gravity-controlled creep of material and the violent, unregulated movements resulting in mudflows and landslides are so hazardous that much time and money have to be spent on preventing or repairing the damage they cause. Landslide repairs to federal-aid highways in the USA cost $43 million in the single year 1973.

In almost all mass movements controlled by gravity, water plays a key role by reducing the strength and coherence of unconsolidated materials or by lubricating potential slip surfaces. Most of the catastrophic flows and landslides of historic times have followed storms or periods of heavy rain. Other factors making for instability are the destruction of a plant cover, the oversteepening of slopes (by natural or human agencies) and the effects of vibration due to earthquakes or other causes. The properties of rock and soil (sand, silt, clay and other consolidated materials are lumped together as 'soil' in civil engineering

terms) that promote mass movement under gravity include:

(a) high pore fluid pressure;
(b) low cohesion in unconsolidated material, close fracturing (on bedding, joints, schistosity) in solid rock;
(c) impermeable layers which obstruct groundwater flow and provide lubricated slip surfaces;
(d) occurrence of 'slippery' rocks such as clay, evaporite, anthracite;
(e) downslope dip of potential slip planes such as bedding, joints, faults, schistosity.

The prevention of gravity-controlled movements during or after the construction of roads, airstrips, dams and buildings depends to a large extent on the recognition and avoidance of potentially unstable combinations of rock, structure and topographical form during the planning stage.

Setting aside floods, which involve only water, mass movement powered by gravitational forces can be classified as follows:

(a) Slow downslope movements taking place by creep.
(b) Rapid downslope movements: (i) flows in which the moving material is internally deformed, and (ii) slides in which material moves more or less en bloc along a basal detachment surface.
(c) Subsidence and bulging in response to loading and unloading.

Slow downslope movements Creep of soil, glacial deposits, weathered or fragmented rock and other incoherent surface materials continues over long periods especially where these materials are subjected to alternate phases of saturation and drying or freezing and thawing (Fig. 6.2a). Creep (or soil creep) is the general term for these movements; solifluction is a term generally applied to movements in permafrost regions (Section 6.2.3). Prolonged downslope creep tilts trees,

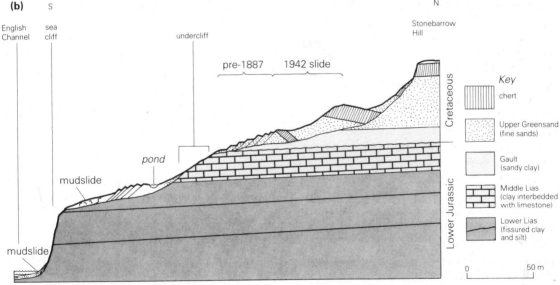

Figure 6.2 Mudflows and landslides (a) Mudflow at Headon Hill, Isle of Wight: the mud is derived from weak Tertiary sediments (Crown Copyright reserved). (b) Landslide formation illustrated by the Fairy Dell, Dorset: note the gentle seaward dip of the bedding and the presence of weak clays on which the Upper Greensand has slipped – the rotation of successive slide masses is indicated by the chert bed (after Brunsden, D. and D. K. C. Jones 1976. *Phil Trans R. Soc. Lond.* **283**).

walls, fences and detached blocks of solid rock which may ultimately topple over. Embankments and retaining walls are bulged and disrupted, and drains, gas or electricity mains and other services broken. Creep may be slowed by improvement of drainage and, in some instances, by the establishment of a grass or tree cover.

Rapid mass movements Soil, mud and debris flows move suddenly and rapidly as jumbled incoherent bodies capable of carrying with them buildings or large rock masses. Such flows have a high (often > 30%) water content and move like viscous fluids, sometimes eroding the underlying surface. A build-up of water pressure in pore spaces after rain or during a thaw may initiate the flow by reducing the cohesion of particles to such an extent that local liquefaction takes place. Vibration due to earthquakes, rockfalls or blasting may provide the final impetus (pile driving is thought to have triggered a flow into the Göta river, Sweden in 1950).

Volcanic mudflows, for which the Indonesian word *lahar* is used, are common accompaniments of sub-aerial eruptions. Pyroclastic material built up rapidly during explosive eruptions becomes unstable when saturated by rain, by melted snow or by water expelled from a crater lake. The explosive eruption in 1980 of Mt St Helens (western USA) triggered several mudflows; and a prehistoric flow from Mt Rainier in the same region covered over 300 km^2 of what has since become a populous region. The AD 79 eruption of Vesuvius gave rise to a flow which engulfed the town of Herculaneum (some 7 km from the crater) in a slurry up to 20 m in thickness.

Sensitive clay (quick clay) and quicksand are sediments in which the detrital particles have an unusually loose packing and consequently a high water content. These materials liquefy or become 'quick' in response to small stimuli and can flow on slopes of only a few degrees. Marine intercalations of sensitive clays in Recent fluvioglacial or periglacial deposits have proved troublesome in both Scandinavia and Canada. In the St Lawrence lowlands east of Ottawa, for example, the Leda clay (deposited during a short-lived marine incursion) has flowed repeatedly, destroying fields, blocking roads and weakening the foundations of buildings. The sites of past flows are marked by many low arcuate scarps.

Debris flows from mine tips and dumps are hazardous in mining and industrial areas where unstable slopes and adverse structural and hydrogeological features in the bedrock promote instability. In the coalfields of South Wales, where pervious sandstones are confined by mudstones in the Upper Carboniferous Coal Measures, mine dumps have been built up over several decades on the steep sides of narrow valleys. The disastrous flow of 1966 which destroyed the village school at Aberfan (South Wales) with the loss of more than 140 lives was derived from a 40 m high tip sited near a spring line related to a mudstone outcrop. During a period of heavy autumn rain, emerging water sapped the tip base until a moment came when the saturated downslope toe suddenly liquefied and poured down into the valley, carrying with it a mass of drier debris from the main body of the tip. The combination of adverse circumstances, recognised only after the event, illustrates the need for thorough site investigation before a new tip or other loading structure is built.

Landslides, landslips and slumps in which material is transported largely *en masse* may affect areas many square kilometres in extent. The detached slide moves on one or more slip surfaces, leaving behind an incised, often arcuate, scar and commonly rotating in such a way that bedding or other identifiable surfaces are tilted back towards the source (Fig. 6.2b). The characteristic scars, the tilting of bedding and other surfaces and the jumbled, poorly drained hillocks formed by old slide masses serve to identify landslide terrains and provide warnings of instability that have not always been heeded by site engineers (Section 6.3.1).

Landsliding is favoured by steep slopes, such as those of immature, rapidly eroded mountain regions and of sea cliffs as well as by the occurrence of structural planes inclined down slope. Natural or artificial over-steepening of gradients – whether by the downcutting of a river, the undercutting of a cliff or the construction of a road or railway – provokes repeated slides and, although drainage and other preventive measures may improve conditions, sliding can seldom be prevented while the fundamental instability persists. Individual slides may be triggered by the build-up of ground water or by external shocks.

Where inclined structural surfaces are present, and especially where they are lubricated by water held up at impervious layers or by slippery clay, fault gouge etc., the moving mass tends to slide preferentially along them (Fig. 6.2). Elsewhere, new slide surfaces may develop with a characteristic spoon-shaped profile. Sliding into narrow valleys may be followed by catastrophic floods owing to the obstruction of rivers or infilling of lakes. These effects are illustrated by the Vaiont slide (northern Italy) described in Section 6.3.1.

Submarine slides affecting large areas of the sea floor take place especially in unstable marine basins where sediment has accumulated rapidly. Sliding of soft sediment without complete loss of coherence gives rise to slump sheets common in many sequences of turbidites and other poorly sorted psammites; loss of coherence leads to the generation of turbidity currents from which turbidites settle out. Both types of flow, often triggered by storms or earthquakes, can damage submarine cables and other installations and are potential dangers in offshore oilfields.

Subsidence and bulging Subsidence and bulging in response to loading, unloading or withdrawal of support are gravity-controlled processes which may take place naturally in regions underlain by weak rocks such as clays and evaporites and which are also common in regions of mining, quarrying and building operations. Natural examples of the effects of unloading in Britain are the valley bulges developed in Northamptonshire and Lincolnshire where Liassic clays overlain by sandstones, limestones and ironstones have risen in response to the removal of overburden by river erosion to form narrow ridges faithfully following the course of a modern valley for as much as 15 km (Fig. 6.3). The hard overlying strata (which include iron ores formerly of considerable economic importance: Section 4.4.4) tend to droop towards the valley floor (cambering) and blocks detached along joints are gradually displaced down slope.

On a smaller scale, the removal of overburden during quarrying and mining may cause weak material to bulge up in the floor of an excavation or tunnel, especially where the water table is high. Upheaval may produce a bulge several metres high. Loading by the construction of a building or other large structure may lead to outward flow of clay or other mobile material in the substratum and hence to subsidence of the foundations. Such movements following the building of the Shell Centre on the south bank of the River Thames (London) at one time threatened damage to the underground railway tunnels beneath. The removal of support from below, either naturally by solution of limestone or salt or by the construction of tunnels or mines is another cause of subsidence. Effects of this sort are common in old-established coal-fields (cf. Fig. 6.1).

Subsidence may also follow the abstraction of fluids – water, oil or gas – where this process is accompanied by compaction. At Long Beach, California, pumping from the Wilmington oilfield induced a surface subsidence of up to 10 m in 30 years. In the vicinity of the port and naval dockyard, wharves,

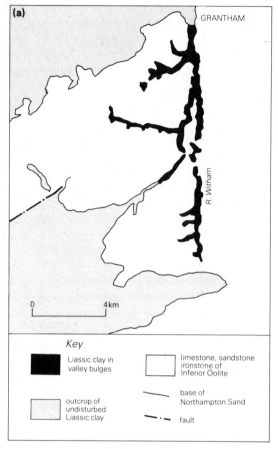

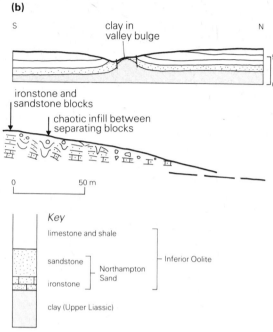

Key

Liassic clay in valley bulges	limestone, sandstone ironstone of Inferior Oolite
outcrop of undisturbed Liassic clay	base of Northampton Sand
	fault

Figure 6.3 Valley bulges (a) The map shows the distribution of Liassic clay bulges on the course of the River Witham, resulting from the removal of the load on weak clays. (b) The sections show a simple bulge uptilting the overlying stronger beds (above), and the effects of cambering in ironstone and sandstone on a valley side (below) – blocks of the strong beds slide down slope, pulling open gulls filled by collapse debris (after Hollingworth, S. E. 1944. *Q. J. Geol Soc. Lond.* **100**).

bridges and roads had to be raised to counteract the effects of compaction, removal of ground water and loading by new buildings. Since 1958, the injection of water into the reservoir rocks has brought subsidence almost to a halt by raising fluid pressures. This example illustrates the importance of taking geological factors into account rather than simply imposing engineering solutions.

6.2.2 Deep mines

The mining of valuable metals, notably the mining of gold in the Witwatersrand basin (Section 4.4.2), may be extended to depths of 2–3 kilometres. At such depths, the mine shafts and levels which form large voids in solid rock are inherently unstable. The mine is preserved from collapse under the prevailing high hydrostatic pressures only by the strength and rigidity of the enclosing rocks. Discontinuities such as faults, and weak zones where the rocks are thinly bedded, heavily fractured or of low strength, provide hazards during the construction of the mine (see below) and tunnels may be lined with concrete, grouted (Section 6.3.1) or supported to prevent subsequent distortion or collapse of the roof and walls. Strong rocks such as quartzite or dolerite may be rendered unstable by the changes in internal stresses brought about by construction of the mine. Rock bursts, during which large volumes of broken rock are projected into the mine with explosive violence, result from sudden fragmentation of strong rocks under the new stress system. Rock bursts of this type are a particular hazard in the deep mines of the Witwatersrand where the gold-bearing conglomerates are enclosed in Precambrian quartzites cut by basic dykes of differing mechanical properties (Fig. 6.4).

Water and gas (including inflammable methane and other hydrocarbons) tend to migrate into all mine passages, which can be kept clear only by the installation of complex arrangements for pumping and ventilation. As deep mines are below the water table, the enclosing rocks are normally saturated and serious flooding may be expected when the advancing mine face penetrates a fault or fracture zone with high hydraulic conductivity (Section 2.4). The high hydrostatic pressure may cause water held in such a zone ahead of, or to one side of, the passage to burst its way into the mine through a screen of normal rock. Conditions leading to catastrophic flooding of this kind are found in the Witwatersrand basin where lateral flow of ground water in cavernous dolomites overlying the ore-bearing Witwatersrand group is obstructed by vertical dolerite dykes which may channel water down into mines in the underlying strata (cf. Fig. 6.3). The risks of flooding and of initiating rockfalls make it essential for the mine geologist to be able to predict the positions of zones of weakness and of water-bearing zones when extensions are being planned. Accurate predictions depend on a detailed reconstruction of the local structure in three dimensions based on geological and geophysical observations at the surface, in cores and in the existing mine levels. The geophysical characteristics of water-filled fracture zones (e.g. low resistivity) may enable such zones to be detected before they are breached by excavation (see Ch. 8).

6.2.3 Permafrost

In polar, sub-arctic and mountain regions, where air temperatures remain at or below 0 °C for long periods, perennially frozen ground or permafrost is developed. Permafrost areas, which occupy some 20% of the world's lands, present special problems not only for the construction and maintenance of buildings, roads and other structures but also for the provision of water and the disposal of waste. Industrial developments in Siberia and North America over the past few decades, and the recent discovery of Arctic oilfields, have focused attention on the need to understand the permafrost regime.

The permafrost layer in which water in voids is normally frozen ranges from less than a metre to several hundred metres in thickness. It is overlain by a suprapermafrost layer (for which the Eskimo word *talik* is sometimes used) in which freezing and thawing

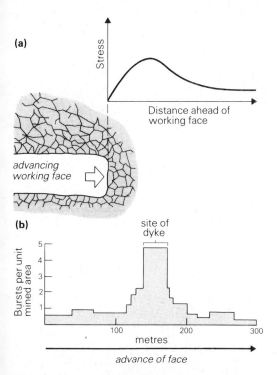

Figure 6.4 Stress effects in underground mines (a) Fracturing and the build-up of stress concentration around an advancing stope. (b) The frequency of rock bursts in gold mines of the Witwatersrand basin in relation to the occurrence of basic dykes.

alternate with the seasons. Temperature conditions are delicately balanced in all but the coldest climates and comparatively small disturbances may bring irreversible changes resulting from the melting of the permafrost layer.

The special features of permafrost regions arise from two principal causes. First, the winter freezing of water in the talik leads to the growth of large ice masses just below the surface which disrupt the soil. Swarms of small hillocks (pingoes) are pushed up above growing ice masses and collapse to waterlogged depressions when the ice melts. Repeated freeze and thaw result in the development of 'patterned ground' in which pebbles and boulders are segregated into reticulate patterns that pass, on inclined surfaces, into stripes and garlands distorted by solifluction. Secondly, the presence of the permafrost layer restricts the movement of ground water during the summer thaw. Water is therefore trapped in the top few metres of soil which becomes weak and prone to **solifluction** (Section 6.2.1).

The drastic loss of strength that follows the spring thaw renders many roads, railways and airstrips unusable in summer. Solifluction, taking place on slopes of no more than a few degrees, may undermine the foundations of buildings, while frost heave – the upward displacement of soil and other materials resulting from the winter growth of ground ice – disrupts roads and may force up or even eject piles used to support buildings and bridges. The first necessity for construction in permafrost regions is to select the sites least prone to these disturbances – that is, sites on solid rock or on coarse-grained, well drained gravels or sand – and to avoid the susceptible clays and other fine-grained materials.

Construction works of all kinds commonly have the additional side effect of disturbing the thermal equilibrium in the soil (Fig. 6.5). Stripping of an insulating vegetation cover prior to construction of a road or airstrip, or compaction resulting from the passage of heavy vehicles, promotes thawing in summer to an extent which may ultimately destroy the permafrost layer. Seasonal temperature fluctuations three to four times as great as those in undisturbed ground were recorded, for example, beneath a road built in the 1950s in Alaska. This road, and many others like it, was severely distorted by settling and frost heave. Modern roads are usually built on a thick bed of 'fill' which helps to insulate the permafrost layer. Construction and maintenance costs tend to be at least twice those in temperate regions.

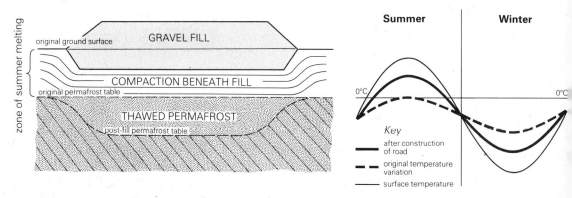

Figure 6.5 Stability in permafrost regions Disturbance of the thermal regime caused by construction work: the top of the permafrost layer is lowered during the summer thaw and the foundations of the road become unstable. A thicker fill of gravel is needed to insulate the permafrost layer and to maintain stability (based on Ferrians, O. J. *et al.* 1969. *USGS Professional Paper* no. 678).

Further disturbances of the thermal regime are found beneath buildings or other large structures set directly on the ground. Heat loss to the soil reduces or destroys the permafrost layer and promotes the inflow of water which subsequently freezes into ground ice. Large heated buildings built in the 1950s in Barrow, Alaska, induced temperature rises of 5–6 °C at depths of 6–7 m and were subject to uneven settling and fracturing. Modern buildings and pipelines are usually raised on stilts, and cold air is circulated through their foundations in order to maintain stability.

6.2.4 Seismic regions

The problems of maintaining stability in seismically or volcanically active terrains need little emphasis. The principal seismic regions and most of the world's active volcanoes are located at plate boundaries where adjacent slabs of the Earth's crust move past each other at rates averaging a few centimetres per year (Fig. 6.6). These persistent movements take place partly by continuous slow creep and partly by occasional abrupt displacements of up to a few metres on discrete faults. The initial effects of such earthquakes are due to the wave-like vibrations sent out from the focus, which achieve accelerations equal to the gravitational acceleration of the Earth. Collapse of buildings and severing of gas, water, sewage and electricity mains may follow. Extensive flooding of coastal regions is caused by the generation of 'tidal waves' (tsunamis) and of valleys by the emptying of reservoirs and lakes. The outbreak of fires, disruption of communications, contamination of water and spread of disease are common secondary consequences.

The heavy casualties and damage resulting from earthquakes in populous regions

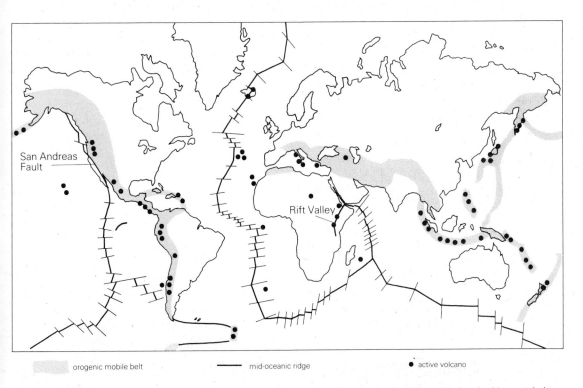

Figure 6.6 Seismic and volcanic hazards The schematic map shows the seismic zones associated with crustal plate boundaries and the positions of major volcanic centres.

93

(143 000 people died, for example, in the Tokyo earthquake of 1923) demonstrate the need to minimise losses both by the adoption of earthquake-resistant designs and by the development of methods of identifying high-risk sites and predicting future shocks. The obvious undesirability of siting dams and public buildings such as hospitals close to active faults makes it unnecessary to labour the need for accurate geological mapping of seismic regions. Yet maps are of limited value unless one understands the processes at work. Even in the intensively studied San Andreas fault zone of California, faults previously thought to be inactive moved during an earthquake in 1971. The almost accidental discovery that changes of fluid pressure in a fault zone near Denver (Colorado) influenced the scale and frequency of earthquakes shows how complex are the factors involved. This discovery in the early 1960s followed the pumping of waste fluids into a deep well in fractured Precambrian basement. The frequency of small shocks was found to correlate with the volume of fluid injected and it was concluded that high fluid pressure facilitated minor (and therefore less destructive) adjustments of rocks under stress.

The possibility of predicting major earthquakes with sufficient accuracy to justify the evacuation of cities or buildings arises mainly from the recognition of deviations from an established pattern of behaviour over periods of months, days or hours before rupture takes place (Table 6.1). Such observations depend in turn on continuous monitoring of the fault zone over a much longer period (see Ch. 8). The remote prospect of earthquake control arises from the possibility that injection of fluids might be used to assist in the gradual release of strain energy, according to the process recognised in the Denver area.

Table 6.1 Earthquake hazards – premonitory signs.

distortion of land surface	Repeated levelling surveys and tiltmeters revealed measurable distortion of the surface over the weeks prior to Japanese earthquakes in 1943 and 1964. Anomalous changes of sea or lake levels have been recorded immediate before earthquakes.
seismicity	Fluctuations in the frequency and scale of small seismic tremors may be systematically related to larger earthquakes. During the period 1965–7 in Japan, for example, intense swarms of small shocks in the Matsushiro area were recorded near the epicentres of earthquakes which took place a few months later.
other geophysical anomalies	Anomalies in the, e.g., magnetic field or electrical conductivity have been recorded.
anomalous groundwater conditions	Unusual flows from springs or seepages and sudden changes of water level in wells have taken place a few days or hours before some earthquakes.
gas seepages	Radon and other gases are commonly emitted from fault zones and changes in the rate of emission may herald earthquakes.
anomalous behaviour of animals	Although difficult to check, evidence of unusual behaviour among many groups of animals is taken seriously, especially in China.

6.2.5 Volcanic regions

The hazards of volcanic activity fall into four main categories:

(a) Pyroclastic flows, most often emitted during the major eruptions of rhyolite volcanoes, are dense, fast moving and generally hot particulate flows which blanket large areas and destroy all life in their path. Although eruptions of Vesuvius (AD 79), Krakatoa (1883) and Mont Pelée (1902) caused severe devastation and loss of life, they do not represent the worst that may be expected from the emission of pyroclastic flows. Vast explosive eruptions taking place at very long intervals have given rise to pyroclastic flows thicker and more extensive than any recorded by human observers – for example, the Taupo ignimbrite of New Zealand (?AD 186) covered an area some 70 km in diameter.

(b) Ashfalls resulting from less violent explosive eruptions (such as that of Mt St Helens in 1980) are more frequent occurrences, though their slower rate of accumulation and small scale make them less hazardous. Even a thin ash layer may, however, destroy vegetation and thick falls destroy buildings and disrupt communications. Lahars, resulting from the flow of saturated pyroclastic ash, carry the same risks as other rapid mudflows (Section 6.2.1).

(c) Lava flows, mainly from basaltic volcanoes which emit lavas with relatively low viscosity capable of travelling some distance from the source, destroy buildings, harbours and roads in their path. Destructive eruptions take place from time to time on the slopes of such volcanoes as Etna (Sicily) and in the volcanic terrains of Iceland and Hawaii, though casualties are generally small.

(d) Flooding, on a small scale due to the emptying of a crater lake or to blocking of a river course, and on a large scale as a result of the displacement of water by pyroclastic flows or ash, is a side effect of volcanic eruptions responsible for many casualties. Most of the 36 000 deaths during the 1883 eruption of Krakatoa were due to drowning by tsunamis sweeping the Sunda Strait between Java and Sumatra after pyroclastic flows entered the sea.

The style and frequency of volcanic eruptions depend on the nature of the magmas supplying them. The relatively quiet and frequent emission of lava that characterises the Hawaiian type of volcanicity, most common in basaltic volcanoes, poses only a limited threat to life and property. The incomparably more violent explosive eruptions of volcanoes fed by viscous magma (especially rhyolitic magma) are so infrequent that we have only an imperfect idea of their place in the cycle of activity at any centre. The extent of the areas at risk during vesuvian or peléan eruptions, and the rapidity with which pyroclastic flows travel, render human efforts to avert damage almost useless. The only practicable means of reducing casualties – evacuation of a threatened area – depends on the development of reliable means of predicting eruptions. Monitoring of individual volcanoes (cf. Ch. 8) has shown that the rise of magma preceding many eruptions leads to distortion of the surface, influx of heat (sometimes observed in the waters of a crater lake), disturbances of the magnetic field and an increase of seismic activity; there remains, as for the prediction of earthquakes, an unbridged gap between the observation of these general premonitory signs and the forecasting of the time and severity of eruption itself.

Table 6.2 Siting of reservoirs.

General considerations

(a) *Catchment area*: rainfall and catchment area must be adequate to maintain water level.

(b) *External hazards*: active seismic and volcanic areas and landslide terrains are generally unsuitable.

(c) *Former land use*: preference is generally given to thinly populated uncultivated sites; reduction of amenities and destruction of wildlife may be offset by landscaping and development of facilities for water sports. Historic sites such as the Nile temples above the Aswan dam may call for expensive conservation works.

(d) *Accessibility*: heavy costs may be involved in the opening up of a remote site.

Technical factors

(a) *Valley morphology*: narrow valleys will store large water volumes with minimum land loss, but valley walls may be unstable.

(b) *Geological structure*: the strength of the dam foundations and stability of valley sides depend largely on geological structure.

(c) *Watertightness*: leakage depends on groundwater conditions and soundness of the dam.

(d) *Construction materials*: local sources of aggregate for the dam are essential.

Figure 6.7 The aftermath of flooding Mud and silt deposits on a Kansas farm (Aerofilms).

6.3 Reservoirs and the control of rivers

6.3.1 Siting and stability

The construction of artificial lakes or reservoirs for water supply, for hydroelectric power schemes and for irrigation and flood control has been among the main concerns of engineering geologists for more than a century (Fig. 6.7). Over 1000 reservoirs are in use in the USA alone. The success of each scheme depends largely on the selection of appropriate sites for the reservoir and, more particularly, for the dam by which water is impounded. As often happens, considerations of convenience, politics, cost or sheer necessity have to be weighed against geological suitability (Table 6.2); thus, for example, many reservoirs have been built to supply water for the large cities of California, despite the evident risks of construction in a highly seismic area.

The principal local factors determining the stability of a dam and the watertightness of a reservoir are those concerning the constitution and structure of the bedrock and the groundwater conditions. Where the dam must be built on weak unconsolidated deposits, its load may be spread by construction of a wide embankment dam in which an impermeable core of concrete or rolled clay is sheathed in a layer of crushed rock and soil. Seepage of water beneath the dam may be reduced by extending the core down through the foundations in a structure known as a cutoff. Weak foundations may be strengthened by preliminary draining and artifici-

Figure 6.8   A concrete dam   Constructed as part of a hydroelectric scheme in Glen Errochty, Scotland (Crown Copyright reserved).

ally induced compaction, by grouting – that is by the injection of a slurry, emulsion or solution of cementing material which will solidify in voids and pore spaces – or by supporting the structure on piles driven into the bedrock.

Where the bedrock is strong, a rigid concrete dam may be built (Fig. 6.8). Faults, major discontinuities and zones of closely spaced parting planes are to be avoided, especially where these structures have a downslope dip. Fractured or cavernous areas may be sealed by grouting.

The accumulation of water behind a dam disturbs both surface and groundwater regimes and initiates reactions tending to restore equilibrium. The water table rises in response to the raising of base level; valley alluvium may flow or compact under the increased weight of water; and erosion may take place at the new shoreline. Where the valley walls are

unstable, these local changes may set off mass movements with catastrophic results. The Vaiont landslide of 1963 which took place in northern Italy soon after the construction of a reservoir in a region of long-term instability may serve as an example (Fig. 6.9). The damming of the Vaiont river, a steep-sided tributary of the Piave river which drains into the Adriatic Sea, was completed in 1960 as part of a hydroelectric scheme. The Jurassic to Cretaceous limestones and clays which underlie the valley are folded into an open syncline so arranged that the bedding planes (and the slippery clay horizons) are inclined towards the valley floor (Fig. 6.9). The inherent weakness of this arrangement is enhanced by heavy jointing due in part to the release of stress caused by rapid river erosion; the scar of an old landslide on the north side of the valley testifies to the instability of the slopes.

97

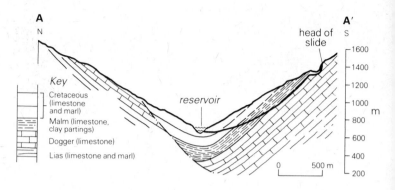

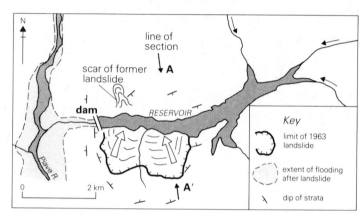

Figure 6.9 The Vaiont disaster The map of the reservoir in northern Italy shows the extent of the landslide of September 1963 and the area flooded by water expelled from the reservoir during this landslide. The section shows the downslope dip of strata and the effects of oversteepening of slopes caused by downcutting of the river in the floor of a broader (glacial) valley (based on Kiersch, A. 1965. *Mineral Information Service* no. 18).

Some minor sliding and creep were observed during and after the building of the dam. The raising of the water table during the filling of the reservoir extended the zone of saturation in the sides of the valley and in mid-September 1963 the rate of creep in the future slide area accelerated from about 1 cm to 30 cm per day. Although this movement was monitored, it was thought at first to be confined to surface materials; not until 8 October, after a week of heavy rain, did observers realise that a large part of the southern slope was sliding as a unit. On 9 October, a rock sheet with a volume of almost 0.25 km^3 detached itself along a slide plane some 100 m below ground and slid bodily downhill at speeds of 25–30 m s^{-1}. Seismic tremors were recorded as far afield as Brussels and, when the slide entered the reservoir, water was expelled with such violence that it leaped the dam and raced down into the Piave valley, destroying several villages and causing the loss of 2600 lives. The dam itself remained unbroken.

Since the waters of a reservoir communicate with the ground water, watertightness depends less on the soundness of the dam than on the hydraulic gradient and the capacity of the bedrock to transmit water. Groundwater flow patterns which can be determined prior to construction (Section 2.5) may indicate the probable extent of leakage and, where necessary, losses may be reduced by grouting of fractured or porous rocks. In general terms, reservoirs whose surface lies below the water table in the flanking slopes will be watertight, whereas those in which levels are above the local water table will lose water to the groundwater system (Fig. 6.10). The first condition is illustrated by the Cow Green reservoir of the Pennines of northern England, which retains water despite the fact that it is flanked by permeable limestones. The second is illustrated by the Aswan reservoir

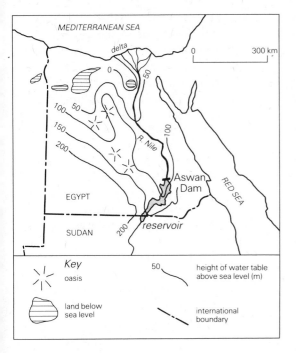

Schematic section

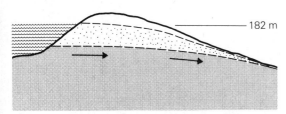

Figure 6.10 The Aswan Dam, Egypt The water table in the Nubian Sandstone slopes northwards towards the Mediterranean and feeds oases where it is intersected by the land surface. In the vicinity of the reservoir above the dam, the water table normally stands at 150–100 m above sea level. Top water level in the reservoir (182 m) is above this height and consequently water may be lost. The coarsely stippled zone in the schematic section shows the possible raising of the water table by this means (based on Knill, J. L. 1971. *Q. J. Engng Geol.* **4**).

on the River Nile which feeds water to the permeable Nubian Sandstone; after the first dam was built in 1902, oases supplied by the Nubian Sandstone further north received an increased discharge for some time (Fig. 6.10).

A long-term problem connected with the management of reservoirs results from the trapping of detritus introduced by the parent river or contributed by slumping from the adjacent slopes. Silting up reduces the water volume and hence the efficiency of reservoirs, especially those of small size. Much of the Nile sediment formerly carried down to Egypt now accumulates behind the Aswan dam. A recent survey in the USA showed that 15% of the reservoirs examined lost more than 3% of their capacity annually. The construction of silt traps up stream, or of outlets near the dam through which silt can be periodically drained away, may reduce the effects of silting up.

6.3.2 River management: The Tennessee River Valley

A bold attempt to raise the prosperity of an area roughly the size of England & Wales by measures that involved the regulation of a whole drainage basin, which was launched in 1933 as part of F. D. Roosevelt's New Deal (USA), provides illustrations of some general points. The Tennessee Valley Authority, set up to administer the scheme, drew funds from both public and private sources and was empowered by Congress to develop an area extending into the states of Tennessee, Kentucky and North Carolina (Fig. 6.11). A fundamental element of the strategy was the construction of many dams and reservoirs along the Tennessee River and its tributaries. These reservoirs were intended to fulfil several functions simultaneously:

(a) *Flood control*: high rainfall in the Appalachian mountains at the head of the basin had been associated with repeated flooding down stream during the decades since the region was opened up. The construction of storage reservoirs on major tributaries, by making it possible for excess water to be held back, has alleviated flood problems as far west as the junction with the Mississippi. In April 1977, almost 40 cm of rain fell in 30 hours in the western Appalachians, an amount more than twice that to be expected for a storm of 100 year recurrence interval. Although there was severe flooding in tributaries of the Tennessee and Ohio Rivers, it was calculated that water levels below the storage reservoirs would have been nearly 4 m higher had the dams not been built.

(b) *Communications*: the raising of water levels behind dams opened up the lower reaches of the Tennessee River to large vessels capable of linking new industrial centres with the Mississippi. Some reservoirs also became recreation centres for sailing and fishing.

(c) *Power supply*: a major objective of the scheme was to assist industrial development in the valley by the provision of hydroelectric power. By the early 1940s,

the TVA had become the second largest supplier of power in the United States and, with the stimulus of wartime needs, production of aluminium, steel, phosphates and chemicals rose sharply.

(d) *Farm management*: reduction of flood damage, regulation of river flow to provide for irrigation, and provision of electricity, combined to allow the redevelopment of lands formerly rendered infertile be soil erosion, overcropping and erratic water supply. The rise in productivity that followed the completion of the scheme was largely a result of improved farm practice, especially the reduction of soil loss by terracing of slopes (Section 7.2.2) and the application of phosphate fertilisers derived

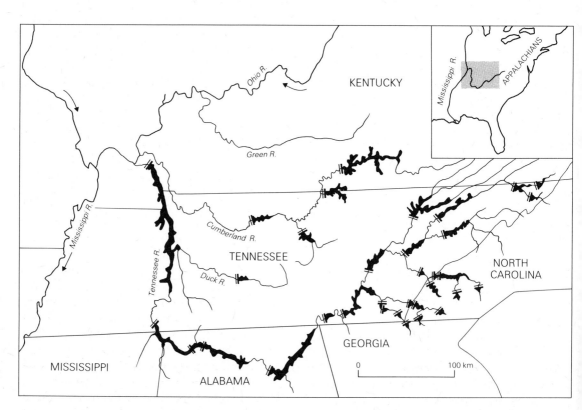

Figure 6.11 River management on a large scale Multiple dams and reservoirs constructed by the Tennessee Valley Authority, USA.

100

from local phosphorite. It will be seen that the river-management aspects of the TVA project, and of similar schemes elsewhere, must be designed as an integral part of a general strategy of development.

6.4 Coastal zones

The beaches, deltas and estuaries where land and sea meet are naturally unstable since they are subject to both erosion and sedimentation and are sensitive to even small changes of relative sea level. The aftermath of the Pleistocene ice age – the period of human development – has been a period of exceptionally rapid fluctuations of sea level which are recorded by drowned settlements and uplifted harbours at many points around the coasts of Europe. An engraving of the 'Temple of Serapis' near Naples (in which the borings of a marine bivalve far above the base of the pillars testify to submergence and subsequent elevation of more than 7 m since Roman times) was used by Charles Lyell as a frontispiece to his *Principles of geology* in 1830 (Fig. 9.5). Long-term geological changes such as these are beyond the power of man to counteract and the geologist is concerned mainly to alleviate the resultant inconveniences by the exercise of foresight. The main themes of such work arise from the need for coast protection, for maintenance of navigation channels and for land reclamation.

6.4.1 *Coast protection and maintenance of navigation channels*
The sea encroaches piecemeal along coasts subject to erosion, by the undercutting of cliffs and removal of beach deposits. These processes are commonly followed by landsliding, as in parts of southern England where slippery Mesozoic clays promote the detachment of interbedded sandstone and limestone (Fig. 6.2). Measures to reduce wave impact or longshore drift by means of sea walls or groynes may be effective at least in the short term. Of more general importance is the protection of low-lying coastal plains from inundation during storms and abnormal high tides. The great deltas such as that of the Ganges in Bangladesh, which lies almost at sea level, are especially vulnerable. In northwest Europe, the storm-prone North Sea has repeatedly inundated the coastal plains of the Netherlands and eastern England where the raising of sea walls and embankments has led over many centuries to the development of largely artificial coastlines. In the Netherlands, coast protection has been combined with land reclamation on a large scale (Section 6.4.2). In Britain, the last catastrophic flood took place in 1953 in the wake of a northerly gale; it was followed by the development of plans to protect London (where some 50 km^2 lie below the possible level of a wind-driven spring tide) by the construction of a barrier capable of closing off the Thames estuary in the vicinity of Woolwich.

Sedimentation in rivers, estuaries and sheltered inlets – the complement of the erosive effects noted above – leads to the silting up of harbours and navigation channels, which require repeated dredging if they are to remain open. Such processes long ago closed ports such as Rye in southern England. The natural effects of alluvial and coastal sedimentation involve the repeated shifting of sandbanks, spits and offshore bars which can be stabilised only in the short term (Fig. 9.2).

6.4.2 *Land reclamation*
The reclamation of land from the sea involves the draining of swampy lagoonal and estuarine areas behind the protection of natural and artificial embankments which keep out the sea. In Britain, cultivatable land has been reclaimed on a fairly modest scale in the Fenlands and the Wash. In the Netherlands, a more ambitious programme continued over some six centuries has involved the enclosure of successive areas (polders) behind protective dykes and the subsequent drying

Figure 6.12 Reclamation from the sea The Noord Oost Polder in the Netherlands (Aerofilms).

out and removal of salt by means of wind- or electrically driven pumps (Fig. 6.12). The scheme scheduled for completion in the mid-1980s which involves the delta complex of channels and islands in the mouths of the Rivers Rhine, Meuse and Scheldt (Fig. 6.13) will serve to illustrate the problems raised by large-scale changes of land use.

Clay, sand and peat have been accumulating in the delta from the late Neogene to the present day in lagoons and swamps behind a protective barrier of coastal dunes. A few tens of metres of Pleistocene to Holocene deposits of these kinds provide the substratum in which new engineering works must be carried out. During the past few centuries, the erection of dykes for flood protection and the damming of creek mouths brought about a gradual coalescence of islands and a simplification of their outlines; and the draining of successive polders allowed much of the land so created to be brought under cultiva-

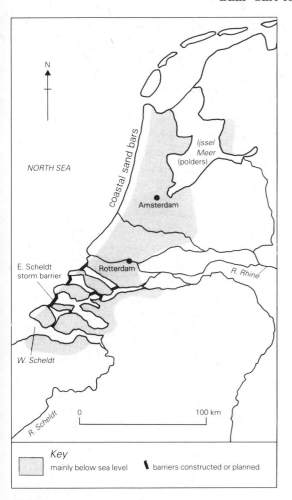

Figure 6.13 Flood control and land use Channel regulation and storm barrier construction in the Scheldt area, Netherlands.

which are the New Waterway and Western Scheldt giving access to Rotterdam (Europe's largest port) and to Antwerp across the frontier in Belgium. The Eastern Scheldt was originally scheduled for closure, but as it is an area of great beauty and it contains old-established oyster and mussel beds, this part of the programme roused public concern. After some debate, a new (expensive) scheme for its preservation as a tidal area protected by a storm barrier capable of being closed in emergency was agreed on. Other tidal channels scheduled for conversion to fresh water will serve as reservoirs for agricultural purposes and perhaps ultimately for much needed water supply.

The geological implications of these changes, affecting an area well over 1000 km^2, have been anticipated as far as possible by modelling, though their working out in practice remains to be tested. The drastic simplification of the coastline (which will be shortened by 600 km) may affect marine erosion and deposition along the outer barriers. The channelling of river waters into a few simplified passages may affect both the deposition of river sediment and the processes of land drainage. Lastly, the conversion of brackish channels and lagoons to freshwater lakes and the consequent elimination of a tidal rise and fall may influence groundwater conditions. Thus, changes begun for social reasons may be expected to initiate further reactions in the natural regime.

6.5 Deep-sea resources

Until recently, neither the composition nor the structure of the oceanic crust was known in enough detail to justify attempts to assess its resources. Several decades of exploration involving geophysical surveys and sampling by means of dredges, corers and deep drilling have shown that the ocean basins have potentially valuable accumulations of fuels and metals and are potential repositories for toxic

tion. Periodic incursions of the North Sea causing heavy damage have stimulated the raising and strengthening of dykes at the sides of the remaining tidal channels.

The great North Sea storm of 1953 was followed by the formulation of a plan to exclude the sea from most of the area, closing the mouths of all but a few channels by dams built roughly in line with the existing coastal dune tract. The waters of the Rhine and other rivers, previously distributed through many tortuous channels, will ultimately be confined to a few strictly regulated routes among waste. The exploitation of resources from the oceans will depend as much on advances in two non-geological fields – those concerned with the 'Law of the Sea' and with the technology of engineering and mining in deep water – as on the success of mineral exploration.

The ownership of resources occurring on or beneath the sea bed has been the subject of legal discussion for decades. The tradition that the high seas beyond the limits of territorial waters (themselves not universally agreed on) are open to all nations is a very old one. An international convention of 1958, while maintaining this freedom, gave coastal states the right to exploit the resources of adjacent parts of the continental shelf. Negotiations under United Nations auspices dealing with the ocean basins proper had not at the time of writing brought agreement on a legal framework within which development could take place.

The expense of operations in deep water makes it inevitable that attention should be focused primarily on the discovery of deposits which are either very large or rich in highly priced commodities. Setting aside placer deposits, most of which are near shore (Section 4.4.2), the most likely objectives are oil and gas in thick sedimentary piles located near continental margins or microcontinents and metalliferous sediments or manganese crusts or nodules carrying high concentrations of nickel, zinc, copper, etc. (Section 4.4.5).

Exploration for such deposits calls for the modification of many standard techniques in order to overcome the problems of working in waters up to 5 km in depth. A first essential is the availability of ships equipped to maintain station while exploration is in progress and to fix the position of a site with extreme precision. Observation of rocks and structures by divers is limited to shallow sites; elsewhere, manned submersibles and/or underwater cameras provide a direct if incomplete view, and sidescan sonar gives a more comprehensive record of sea-floor topography. Dredging, coring and deep drilling provide samples for petrographical and geochemical analysis, and geophysical techniques – especially seismic and magnetometer surveys – give pictures of the structure at depth (see Ch. 8). Drilling techniques developed during the international Deep Sea Drilling Programme (DSDP) have advanced to the stage at which routine exploration can proceed in water depths of 3–4 km.

The possible distribution of economic quantities of hydrocarbons depends on that of sedimentary accumulations thick enough to have allowed the maturation of sapropel to reach an appropriate stage. As Figure 6.14 shows, the floors of the open oceans are generally covered only by a thin skin of sediment. Thick prisms which represent possible targets for exploration are situated at passive continental margins and in narrow tracts such as the Rockall Trough located between a continent and a microcontinent.

Development of and production from oilfields in oceans too deep to allow the use of platforms standing on the sea bed would probably be based on floating platforms linked to the wellhead by flexible flowlines. The mining of metalliferous deposits presents additional problems because manganese crusts and nodules – the most likely targets – tend to form layers only one or two nodules thick which would have to be removed from a large area in order to maintain production. A nodule abundance of 10 kg m^{-2} and an average Ni + Cu of 2–3%, at a site where weather conditions would allow mining on about 300 days per annum, are thought to offer conditions for successful operation – conditions which could be met at several sites in the Pacific off Central and South America. Recovery of nodules could be achieved either by mechanical means, using robust buckets spaced along a cable slung between two ships, or by hydraulic devices using water and compressed air to suck up and raise the nodules.

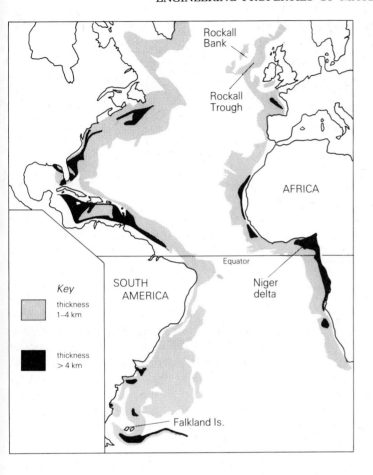

Figure 6.14 Ocean sediments Sediment thicknesses in the Atlantic Ocean, showing the distribution of thick accumulations that include potential sources of hydrocarbons.

6.6 Engineering properties of geological materials

It will already have become apparent that the material properties of most interest to engineering geologists are those that influence the ways in which rocks respond to changes in near surface environments. Among the important properties of materials to be used as foundations for buildings and roads, or in the building of embankments and embankment dams, are permeability, strength, compressibility and chemical stability. The site investigation carried out when a dam or other large structure is to be constructed involves the assessment of these properties both by means of laboratory tests on representative samples or drill cores and by *in-situ* tests designed to reveal the effects of local inhomogeneities and discontinuities. Some of the methods used in site investigation are outlined in Chapter 8; the remainder of this chapter covers points specifically concerned with engineering properties.

The behaviour of unconsolidated materials (the 'soils' of engineering terminology) is the province of soil mechanics. Grain size, sorting and fabric determine permeability and resistance to loading and shearing deformation (see below). Table 6.3 shows that gravels and sands are suitable for many engineering needs and these materials are consequently extracted on a very large scale in all developed countries (Ch. 5). Fine silts, clays

Table 6.3 Engineering properties of unconsolidated materials.

	Permeability	Compressibility	Shearing strength	Suitable for:		
				Foundations	Road fill	Embankments, embankment dams
Pebbly						
clean gravels	very high	very low	high	✓	✓	outer parts
gravels with interstitial fines	moderate	very low	high to moderate	✓	(✓)	core
Sandy						
clean sands	very high	very low	high	✓	✓	outer parts
sands with interstitial fines	moderate to low	low	high to moderate	(✓)		core
Silts and clays						
silts and silty clays	low	moderate	low			core
plastic clays	very low	high	very low			
soils with high humus content	usually low	very high	very low			

✓ = suitable; (✓) = second best.

and materials rich in organic matter have few uses because they are weak, often plastic, and respond to variations in water content by swelling or shrinking. The clay mineral montmorillonite and bentonitic clays derived from weathered volcanics show particularly large volume changes. Clays that have been deeply buried sometimes swell in response to release of pressure (over-consolidated clays). Soluble or chemically unstable components such as salt, gypsum, sulphides and organic matter have deleterious effects on materials of all size grades.

The mechanical properties and structure of lithified and crystalline rocks determine their value as foundations for dams and other structures and are important in relation to the layout of quarries and the construction of tunnels and mines. The inherent variations of such rocks with respect to stresses of various types (Table 6.4) are often of less practical concern than the degree of weathering and the spacing and arrangement of natural fractures. Heavily fractured and weathered rocks are potential hazards not only because their reduced strength renders them liable to failure but also because they may act as conduits for ground water and so cause flooding in

mines and tunnels. A rough and ready measure of the closeness of fracturing is given by the way in which a drill core breaks up during recovery. The **rock quality designation**

Table 6.4 Engineering properties of lithified and crystalline rocks.

Quality of rock controlled by:

uniaxial compressive strength	measured by loading an unconfined rock cylinder to the point of failure: strong rocks require loads >500 kg cm^{-2}, weak rocks <50 kg cm^{-2} (cf. unconsolidated material <2 kg cm^{-2})
shear strength	measured in a shear box, the upper and lower parts of which can be displaced horizontally relative to each other

Compressive and shear strength depend on:

lithology	massive rocks with interlocking grain boundaries have high RQD
weathering state	fresh, moderately weathered, highly weathered (decreasing RQD)
bedding characteristics	massive, thickly bedded, thinly bedded, laminated (decreasing RQD)

Rock quality designation (RQD) = % of core recovered in sticks >10 cm. 100–75% good, 75–50% fair, <50% poor.

(RQD) is expressed as the percentage of the core which is recovered in sticks longer than 10 cm; high-quality materials have RQDs greater than 75%.

The importance of preliminary investigations *in situ* arises from the need to identify rock types and to establish the structure in three dimensions. Detailed geological mapping is supplemented by the drilling of short cores and by geophysical (seismic and resistivity) surveys designed to reveal such features as the depth to bedrock or the thickness of a weathered mantle. A dam site, for example, would preferably be underlain by strong rocks, without intercalations of other materials or zones of close fracturing, and free from inclined discontinuities which might promote slip. The layout of a quarry is often dictated by the need to ensure stability of the working face and (where building stone is to be quarried) to take advantage of rectangular joint systems. In mines and tunnels, zones of weak or fractured rock are dangerous to excavate and may require permanent support. Residual stresses in bedrock may be gradually relieved by distortion of the walls or may give rise to catastrophic rock bursts (Fig. 6.4).

References

Blyth, F. G. H. and M. de Freitas 1974. *A geology for engineers.* London: Edward Arnold.

Francis, P. W. 1976. *Volcanoes.* London: Penguin.

Geological Society of London 1972. The preparation of maps and plans in terms of Engineering Geology. *Q. J. Engng Geol.* **5** (4).

Howard, A. D. and I. Remson 1978. *Geology in environmental planning.* New York: McGraw-Hill.

Knill, J. L. (ed.) 1978. *Industrial geology.* Oxford: Oxford University Press.

McLean, A. C. and C. D. Gribble 1979. *Geology for civil engineers.* London: George Allen & Unwin.

Obert, L. and W. I. Duvale 1967. *Rock mechanics and the design of structures in rock.* New York: Wiley.

Open University 1976. *Urban geology.* S333. Milton Keynes: The Open University.

Tank. R. W. (ed.) 1973. *Focus on environmental geology.* Oxford: Oxford University Press.

7 Interaction with the biosphere

7.1 Geological processes in the biosphere

The chemically active surface zone of the Earth, in which solid rock reacts with water and air, is the natural habitat of almost all forms of life. In this **biosphere**, the activities of animals and plants modify the inorganic processes of erosion, weathering and sedimentation in many ways. The balance maintained between plant or animal communities and their environments is unstable because it can be upset by changes in the climatic, geomorphological and geological controls on the one hand and by changes due to organic evolution and the migration of species on the other. Man, as a dominant species in many regions, plays a major role in the biosphere not only as an agent of geological change but also by direct intervention in the development of faunas and floras. This chapter is concerned with some aspects of the interaction between life processes and geological processes.

A start is made with the soil, the medium in which plants grow. Fertilisers needed to maintain the fertility of cultivated land are dealt with rather briefly because they have been mentioned already in other connections (Ch. 5). Natural variations in the distribution of trace elements that are either essential for, or injurous to, the health of plants, animals or man are treated in relation to their influence in agriculture and medicine. This subject leads finally to a consideration of the effects that mining and industrial processes may have on the redistribution of toxic minerals and elements, and of problems connected with the disposal of waste.

7.2 Soils

7.2.1 Composition

Soil, which provides the starting point for most land-based food chains, is developed from weathered bedrock or unconsolidated materials by reactions with air and water and by the activity of plants, animals and bacteria (Fig. 7.1). A characteristically high level of biological activity is fostered by the simultaneous availability of oxygen, water and organic matter. A mature soil which has achieved a rough equilibrium may range from a few centimetres to several metres in thickness. Colour, structure and composition depend on climate, topography and bedrock and, in turn, determine the nature of the vegetation. A crude vertical zoning defines the soil profile (see later), but the mixing effects of root growth and burrowing organisms keep the structure more or less massive. The composition of soil can be dealt with in terms of four principal components: mineral constituents, organic matter, water and air.

Mineral constituents Weathering and disintegration of the parent rock supply most of the mineral constituents, which may be augmented by contributions carried by wind and streams. These constituents, generally forming about 50% of the total volume, are principally of sand, silt and clay size grades. Texture and composition depend largely on

Figure 7.1 Soil development Soil developed *in situ* on Jurassic limestone in Worcestershire (Crown Copyright reserved).

the proportions in which these grades are mixed. A sandy soil is loose, light, permeable and easily leached. A clay soil is heavy, sticky when wet, hard when dry, poorly permeable and not easily leached. Silty soils, which are common on alluvial plains and loams in which sand, silt and clay are present in almost equal amounts are light, fertile and easily worked. Soil texture in cultivated lands is modified by digging and ploughing, which help to increase porosity, and by the addition of humus and fertilisers.

Soil type is generally related to the nature of the parent rock, although the evolution of the soil profile also varies according to climate (Section 7.2.2). Thus, for example, granites, sandstones and quartz-rich metamorphic rocks tend to yield barren sandy soils, whereas argillaceous sediments and intermediate or mafic lavas give finer and more fertile soils (cf. Fig. 7.3). Limestones, on which solution weathering is a dominant process, may yield little or no soil other than a thin clay derived from impurities. Ultrabasic rocks and metallic ore deposits give abnormal soils in which certain metals may reach toxic levels (see Section 7.4). Fertile alluvial silts, replenished by deposits from annual floods (cf. Fig. 6.7), provided the basis for ancient civilisations in the Nile delta and in the Middle East. The fine, somewhat calcareous, periglacial loess yields fertile soil in parts of China.

Elements needed for plant growth which are supplied by the mineral constituents (Table 7.2) must be made available to the soil water before they can be absorbed by roots. The **base exchange** property of clay minerals – the capacity to form loose bonds with a variety of cations which can be exchanged for cations in solution – facilitates this process. Clay minerals linked to organic molecules produce negatively charged particles (micelles), capable of absorbing H, Ca, K, Mg, NH_4 and other cations which can be released to solution. The **exchange capacity** of the soil depends on the number of ions held in this way. Bacteria may act as mediators in some of the exchange processes.

Organic matter Organic debris accumulating at the soil surface is worked over by scavengers and worms and is decomposed by bacterial and fungal action. By these means, plant debris is broken down to simpler organic compounds and mixed with the mineral constituents of soil, darkening the colour and modifying the texture. Humus is the general term for this organic matter in soil which constitutes a few per cent of the total. The oxygenating environment (see below) generally ensures ultimate breakdown to H_2O and CO_2. Where soils become anaerobic through poor drainage or a too rapid build-up of humus, bacterial decay may be halted before breakdown is complete (see below).

In the chemically active soil environment, and especially where organic matter accumulates quickly and ground water becomes acid, iron, manganese, cobalt, uranium, vanadium and other metals may be fixed in complex organic compounds, occasionally reaching toxic levels (see Section 7.4). Iron or manganese may be precipitated in nodular or pisolitic segregations (hardpan) as a result of these reactions.

Ground water and air As soils have a high porosity, water and air together form up to about 50% of the total volume. As the majority of soils lie above the water table and are therefore not permanently saturated, the proportion of water in the voids varies with rainfall and evaporation. Where rainfall is moderate, water tends to drain down through the soil. Retention in the soil is favoured by the presence of humus and by fine grain size which increases capillary forces. In arid climates where evaporation is high, soil water may be drawn upward by capillary action and be lost from the surface.

Ground water carries both inorganic anions and cations released by reactions in the soil and organic radicles formed from the decomposing humus. When the soil is well drained, calcium, alkalies and magnesium are carried down towards the water table, their places on the micelle particles being taken by hydrogen ions. This process of **leaching** leaves the upper soil layer depleted in a number of elements needed by the plant. In arid climates, surface evaporation may lead to the concentration of calcium carbonate, sulphates or salt which renders the soil infertile (Table 2.1). After prolonged evaporation hard **duricrusts** may accumulate (calcrete is mainly $CaCO_3$, silcrete SiO_2. The pH (section 2.2) ranges from about 3 in acid soils to about 9 in calcareous soils (pH7 = neutral). The pH controls the nature of the vegetation a soil will support; it can be determined roughly by means of an indicator such as litmus and extremes counteracted by appropriate additions.

Voids not occupied by water are filled with air in free communication with the atmosphere. Soil air is generally saturated with water and slightly richer in CO_2 than atmospheric air. Oxygen (around 20% of the total, as in the atmosphere), is generally sufficient to ensure ultimate breakdown of organic matter and thus to prevent an undue build-up of humic acids which halt bacterial decomposition. Soils depleted in oxygen became grey or black, evil smelling and often infertile (gley soils). Rapid accumulation of organic matter in an oxygen-poor environment leads to the build-up of humic acid, the cessation of bacterial activity, the accumulation of peat and hence to coal formation (Section 3.6). Nitrogen, which forms almost 80% of the atmosphere and is essential to life, cannot be directly absorbed from the soil atmosphere by most plants. Exchangeable nitrogen in the form of ammonia derivatives held on soil colloids is readily oxidised to nitrates and (since most nitrates are soluble) tends to be lost by leaching. Nitrogen-fixing bacteria living symbiotically in the roots of leguminous plants give these plants a value in the maintenance of fertility.

7.2.2 The soil profile and soil types

The soil profile (cf. Fig. 7.1) defines the vertical variations in a soil as seen in section. In general terms, the uppermost soil layer (the A zone), often leached (see above) but enriched in humus, is distinguished from the underlying B zone in which mineral matter is sometimes redeposited and from the C zone which grades into unaltered bedrock.

The effects of groundwater flow on the development of the soil profile are reflected in a close relationship between soil type and climatic conditions (Table 7.1). The regional distribution of the principal soil types therefore corresponds broadly with that of the climatic zones. Smaller-scale variations in soil type are related both to the character of the bedrock (see above) and to the topography. In mountain tracts, vertical climatic variations give rise to corresponding soil variations. Local irregularities such as are due to soil creep or to minor obstructions of drainage may further modify the development of the soil profile.

The climatic control of soil types is most clearly expressed in lowland areas and it is therefore no accident that the accepted terminology is derived from Russian and American authors. As the details given in Table 7.1 indicate, climatic extremes – whether sub-arctic, arid or tropical – tend to give poor soils which deteriorate under intensive cultivation. The world's major crop-producing zones are those characterised by temperate and warm-temperate climates where brown podzols and dark chernozems are developed. In these soils, the rate of bacterial decay and mineral reaction is sufficient to maintain a supply of nutrients; leaching of the A zone is not intense and hardpan does not build up rapidly in the B zone. Fertility can be sustained by the addition of fertilisers, the nutrients of which are converted to available forms by the activity of the flourishing soil communities.

Table 7.1 Soil types in relation to climate.

tundra soils (sub-arctic)	Formed largely in areas of permafrost (Section 6.2.3). Impeded drainage and instability resulting from summer thaw in suprapermafrost layer prevent development of well defined profile. Soils generally thin, often waterlogged and anaerobic (gleys). Flora stunted.
podzols	Formed in cool humid climates. The A zone is strongly leached and is consequently white or pale grey (Russian *zola*, ash) the B layer may contain hardpan with Fe, Mn, Co etc. Low average temperatures slow down bacterial decomposition and favour acid conditions (owing to build-up of humic acid), with accumulation of peat above the A horizon. Mainly conifers.
grey-brown podzols (brown earth)	Formed in temperate humid climates. Less severe leaching of A zone than in podzols, and more rapid decay and mixing of humus give a darker and more fertile A zone. Mainly deciduous forest.
chernozem (black earth)	Formed in temperate subhumid climates. Little or no leaching due to low rainfall, carbonate deposits in B zone. Deep, fertile soils, vulnerable to soil erosion. Mainly grassland.
desert soils	Formed under arid conditions. Thin soils with low content of humus, oxidation gives red-brown colour, salts in surface layers, sometimes duricrusts. Sparse scrub.
tropical and subtropical soils	Formed in hot humid conditions. Rapid decomposition and strong leaching give an A zone which tends to lose its fertility under cultivation and may be lateritised. Rain forest.

7.3 Management of soils

7.3.1 Stability

In a mature system, the living plant cover draws on and contributes to the underlying soil to maintain a rough balance. This balance is subject to the natural effects of climatic or geological change and, more drastically, to modifications by human agencies. Two aspects of the management of soils call for brief comment: first, the maintenance of physical stability; secondly, the supply of nutrients.

A soil is stabilised by its plant cover which binds particles into a coherent mat, slows down runoff of rain and contributes moisture-retaining humus. Loss of the plant cover – which may take place naturally in drought, or after floods or fires – is generally followed by a speeding up of hill creep and erosion by streams and wind. Overgrazing by domestic animals, felling of trees and cultivation of crops commonly lead to exposure of the soil and resultant erosion. In northern Mississippi in the 1960s, loss of sediment from land under cultivation (as measured by the sediment load of rivers) was found to be almost three orders of magnitude greater than that from mature forest land. A disastrous drought in the 'dust-bowl' of central USA during the depression years of 1933–4 was accompanied by dust storms which stripped the soil from dried-out fields and brought wheat production to a halt over much of Oklahoma.

Soil erosion in the broad sense commonly involves not only the fertile topsoil but also the weathered bedrock exposed by breaching of the soil. In semi-arid climates, such as that of the western interior of America, deflation by wind is accompanied by gullying after heavy rain, giving a dissected badland almost useless for cultivation. Reclamation of land damaged by deflation, or of migrating sand dunes, depends on the establishment of a cover of fast-growing plants which bind soil particles and help to conserve moisture. In mountain terrains, soil loss by hill wash and hill creep is speeded up by destruction of the plant cover. In the Himalayan state of Nepal, soil erosion has increased over the past decades in line with a rise in population; the more rapid cutting of firewood – the principal fuel – and the extension of cultivation needed to provide for larger numbers of people have speeded up losses of topsoil which ultimately finds its way into the Bay of Bengal. The traditional method of reducing losses is by

Table 7.2 Fertilisers.

Principal nutrient	Source	References
nitrogen N	organic fertilisers including animal manure, plant waste, seaweed, fishmeal, dried blood etc, give exchangable nitrogen	
	nitrogen-fixing bacteria in soil or in symbiotic relationship with legumes	
	nitrates, non-marine evaporites of Chile	Table 5.4
	ammonia and its derivatives from petrochemical plants	Table 5.3
potash K	organic fertilisers, especially manure	
	bittern salts of marine evaporites	Table 5.4
lime Ca	shells and bone	
	carbonate rocks including limestone, tufa, calcrete	
phosphates P	bone meal, manure	
	guano (consolidated droppings of sea birds, from oceanic islands, now largely worked out)	
	superphosphate derived from phosphorite	Section 5.6
	basic slag byproduct of steel production	
sulphur S	salt domes (reduction of gypsum)	Table 5.4
	iron sulphides	

Figure 7.2 Terrace cultivation on steep slopes The photo shows old terraces below the Crusaders' castle of Le Krak de Chevalier, Lebanon (Aerofilms).

contour terracing and the provision of drainage channels for surplus water (Fig. 7.2).

7.3.2 Fertilisers

Under natural conditions, the cycle of growth and decay *in situ* preserves the chemical balance of the soil. The harvesting of crops – whether of cereals, sugar, lumber, wool or beef – involves the removal year by year of some of the produce of the land; it may be followed by a decline in fertility. In favoured environments, fertility may be maintained naturally, for example, by rapid weathering of newly erupted volcanic rocks or by the annual deposition of new silt on an alluvial plain (cf. Fig. 6.7) as in the Nile delta prior to construction of dams up stream (Section 6.3.1). More generally, fertilisers are added to make good losses or counteract natural deficiencies. The fertilisers mentioned here (Table 7.2) provide the major elements nitrogen, potassium, calcium, phosphorus and sulphur essential to plant growth. The role of trace elements is touched on in Section 7.4.

Organic fertilisers such as farmyard manure, compost and seaweed have been in use as means of recycling waste matter almost since farming began. Natural inorganic fertilisers are derived from source rocks containing anomalous concentrations of the desired element. With the exception of limestone (in short supply in some metamorphic terrains, especially central Africa, but generally freely available) these rocks are relatively rare and therefore expensive; in the mid-1970s, British farmers spent roughly £300 million annually on fertilisers supplying nitrogen (the principal nutrient), phosphorus and potash. The origins of natural and some synthetic fertilisers have been dealt with in Chapter 5 and the appropriate references are listed in Table 7.2.

Once fertilisers have been added to the soil, their effects depend on the manner in which they enter into the natural soil system. Leaching tends to remove soluble substances from the A zone to which fertilisers are added. Nitrate, sulphate, calcium and magnesium are readily leached from both sandy and silty soils, potash is lost mainly from sandy soils. Phosphorus, on the other hand, may be made unavailable by precipitation in insoluble forms or by adsorption on clay-humus particles. Lastly, fertilisers of any type may be removed wholesale during soil erosion or by sheet flood after heavy rain. Losses due to these processes, and the problems that arise from the transfer of nitrates to the drainage system (Fig. 7.5), may be minimised by matching input of fertiliser to soil type and by timing additions to ensure rapid uptake during periods of active plant growth.

7.4 Geochemical factors in plant, animal and human health

7.4.1 Sources of essential and toxic elements

Almost all the elements available at the Earth's surface occur in trace amounts in plant and animal tissues. The concentration of any element in such tissues is however only indirectly related to that in the source materials, because metabolic reactions operate selectively in the build-up of organic compounds. Many trace elements play crucial roles in one or more biological processes and are therefore essential to health. Many – including some of the essential elements – are toxic in high concentrations. Under natural conditions where plant and animal communities draw their substance, either directly or via a short food chain, from the soil, water and air of a limited area, deficiencies or toxicities may be related to geochemical anomalies in soils and ground water within such an area. Where crops and animals are raised under regimes that involve the use of fertilisers and feedstuffs, a relationship with surface geochemistry is still commonly detectable, as it is in the most nearly self-sufficient human communities. A number of disorders in man that are connected with the deficiency of or excess of certain elements show regional variations of frequency that appear to be connected with variations of element abundances in the natural environment. Regional geochemical surveys based on the sampling of ground water, stream water, soil, stream and lake sediments (Section 8.4) provide the raw data by means of which these empirical relationships can be investigated.

Soil and water compositions vary regionally in relation to climate (which controls weathering style) and locally in relation to bedrock composition (see Section 7.2). The effects of the bedrock are illustrated in Figure 7.3 which compares the relative abundances of some important trace elements in soils developed on granitic and argillaceous rocks in Scotland. Such natural variations are modified in various ways by human intervention (see Section 7.6). The addition of fertilisers to make good deficiencies in nutrients not only changes the soil composition but may also lead to the escape of new substances to the surface and ground water. Mining of coal, metalliferous ore deposits and non-

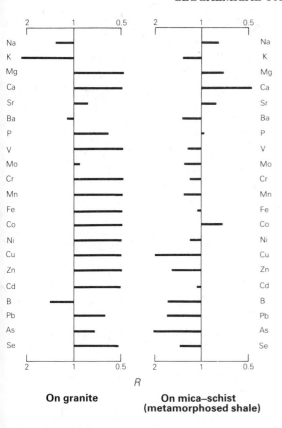

On granite

**On mica–schist
(metamorphosed shale)**

Figure 7.3 Soil composition in relation to bedrock
Variations illustrated by soils on granitic bedrock (left) and on metamorphosed pelitic sediments (right) in Scotland. Relative scales: R=(value in soil tested)/(mean of soils on 10 bedrock types) (based on West, T. S. 1981. *Phil Trans R. Soc. Lond.* B **295**).

substances brought into being by human agencies (the causes of pollution, see Section 7.6) are superimposed on and interact with natural anomalies due to geological factors.

Elements known to be *essential* to the health of plants, animals or man include:

(i) Major elements: C H O N S Ca P K Na Cl Mg

(ii) Trace elements: Fe I Cu Mn Zn Co Mo Se Cr B

Elements known to be *toxic* at abnormally high concentrations include As, Cd, Hg, Pb, Se, U and other radioelements.

7.4.2 Geochemical influences in agriculture and medicine

A long familiarity with the importance of certain elements in maintaining human health is indicated by the fact that, in the treatment of anaemia, the Greeks are said to have used drinking water in which a rusty sword had been steeped (iron is essential for the synthesis of haemoglobin). The variety of trace elements used in the metabolic processes of plants and animals is only now becoming apparent and the paths by which they may be taken up are not yet fully understood. Table 7.3 gives details of deficiencies and toxicities connected with variations in the abundance of trace elements which appear to relate at least in part to the geochemical environment. Some poorly understood relationships (for example, an apparently reduced incidence of cardiovascular disease in hardwater areas) are omitted; disorders related to mineral occurrences are dealt with in Section 7.5.

The complexity of the biogeochemical interactions which control the uptake of essential or toxic elements can be illustrated by reference to three examples.

metalliferous raw materials leads to the build-up of spoil which, on weathering and leaching, releases unusual supplies of certain elements to soils, stream sediments and ground water. Industrial and domestic waste-dumps undergo similar processes (see Table 2.1). Smelters, refineries and factories generate waste products which are dispersed in fumes and effluents (Fig. 7.8). Lastly, the nuclear fission processes involved in the testing of weapons above ground have led to the release of radioisotopes into the atmosphere. Anomalous concentrations of harmful

Table 7.3 Trace elements and health.

Element		Environment
	Too little – deficiencies	
cobalt	wasting and anaemia in grazing animals	on soils on acid igneous rock, sandstone, limestone
copper	wasting, delayed maturity in grazing stock	as for Co; on molybdenum-rich soils
iodine	goitre (defective function of thyroid) in man and livestock	especially in recently glaciated terrains, Alps, Pyrenees, Himalayas, Andes; deficiency mainly due to low levels in water
iron	anaemia due to role of Fe in haemoglobin synthesis	in man, usually associated with dietary or physiological factors
selenium	muscular dystrophy in calves; keshan disease (weakness of heart muscle) in man	sandy or strongly leached soils, occasionally on black earth; central USA, Sichuan province (China), New Zealand
	Too much – toxicity	
arsenic	lethal at high concentrations, plant growth inhibited	on soils contaminated by leaching of sulphide ore bodies and mine dumps, or by industrial effluents (arsenical poisoning mainly from other sources)
cadmium	kidney, bone and liver degeneration	as for arsenic
lead	damage to kidney and nervous system, impaired synthesis of blood protein	as for arsenic, contamination of soft water, industrial products, petrol
mercury	Minamata disease (damage to central nervous system)	as for arsenic; near certain fumaroles and volcanic vents: dietary (high seafood) factor controlled high intake in Minamata Bay, Japan
molybdenum	molybdenum-induced Cu deficiency in grazing animals	mainly on marine black shales
radioactive elements	increased incidence of some forms of cancer, at high concentrations possible genetic damage	usually from non-geological sources: leaching of uranium deposits and mines, diffusion of radon from building materials with high radioactivity locally significant
selenium	in cattle and horses, hair loss, lameness, emaciation; when acute, respiratory diseases, collapse	mainly on marine black shales

(a) *Cobalt* is an essential element for the synthesis of vitamin B_{12} in animals. Deficiency causes slow growth at subclinical levels and leads to anaemia and muscular wasting in acute cases. Co held on clay–humus particles in soil is released rather slowly to plant roots. Stock raised on an area of poorly drained, low-grade pasture in Malaysia showed no sign of cobalt deficiency. When the pasture was upgraded by the introduction of quick growing plant species, however, signs of cobalt deficiency appeared among the stock. In this instance, the more productive plant species proved less effective in transmitting Co to grazing animals, although the underlying soil was not severely deficient.

(b) *Copper* deficiencies, occasionally found on sandy, acid or calcareous soils (cf. Fig. 7.3) lead to reductions in grain yield of cereals and to a variety of symptoms ranging from slow growth and delayed maturity to severe disorders and early death in grazing animals. The symptoms of Cu deficiency in animals often appear in regions where Cu contents are normal but where high molybdenum inhibits the uptake of copper. In Britain, such

116

molybdenum-induced Cu deficiency is recorded on dark Carboniferous and Jurassic shales rich in Mo. The administration of supplementary copper to cattle improves the growth rate of young animals.

(c) *Lead* at high levels is detrimental to human health and is thought to be associated with mental retardation in children. Many areas of sulphide mineralisation show anomalously high Pb in stream sediments and soils, especially in localities such as the Mendip Hills and Pennines in Britain where mining over a long period has exposed sulphide-rich material to leaching. In such areas, Pb values in stream sediments locally rise to 3% (background values 10–150 ppm) and may be associated with other toxic elements such as cadmium, copper, zinc or arsenic. Uptake of these elements from vegetables raised locally (and, by young children, from the direct ingestion of soil) accounts for a few cases of lead toxicity. More common causes, however, are soft water contaminated by the solution of lead from old water pipes, lead paint and exhaust fumes derived from petrol containing lead additives. None of these sources relate to anomalous concentrations of lead in soil or bedrock.

7.5 Minerals injurious to health

Disorders of humans, resulting not from the toxic effects of individual elements but from the inhalation of minute particles of certain minerals, are recorded among the inhabitants of a few anomalous localities and, more seriously, among occupational groups exposed to industrial processes involving these minerals. Ill effects are associated with the intake either of equidimensional particles < 10 μm in diameter or of fibres < 3 μm in diameter in quantities that overload the natural defence mechanisms. Once lodged in the minute air spaces (alveoli) of the lung, these mineral particles stimulate the growth of fibrous tissue and disrupt blood capillaries. The capacity of the alveoli to take up oxygen is permanently impaired and the long-term risk of cancer of the lung or its membranes is increased by the retention of only a few tens of grams of equidimensional grains or a gram or so of fibres.

Respiratory diseases associated with the inhalation of dust, not infrequently culminating in tuberculosis though rarely in cancer, have been common among miners, quarrymen and stonemasons for generations. Silicosis, due to the presence of quartz particles in the appropriate size grade, is the most severe disorder resulting from the intake of equidimensional particles. Diseases associated with mineral fibres are recorded in a dust-prone locality in Turkey where young volcanics used for building contain a fibrous zeolite and, more widely, among those concerned with the mining, processing or use of asbestos (Section 5.2.6). Asbestosis is associated with two principal mineral species. The fibrous serpentine chrysotile forms curly fibres that do not migrate through lung tissue and ultimately disintegrate. The sodic amphiboles crocidolite and amesite ('blue asbestos') form straighter, stiffer and more durable fibres that appear to be associated with raised incidence of cancer in later life. New methods of processing and better ventilation in mines have brought about improvements with respect to silicosis and allied disorders; the risks arising from the use of blue asbestos have only recently become apparent and it seems probable that asbestos will in future have to be replaced by other insulating materials.

7.6 Pollution and waste disposal

The byproducts of human activity – domestic, agricultural, industrial and military – find their way into the air, the surface and ground waters, the soils and other superficial de-

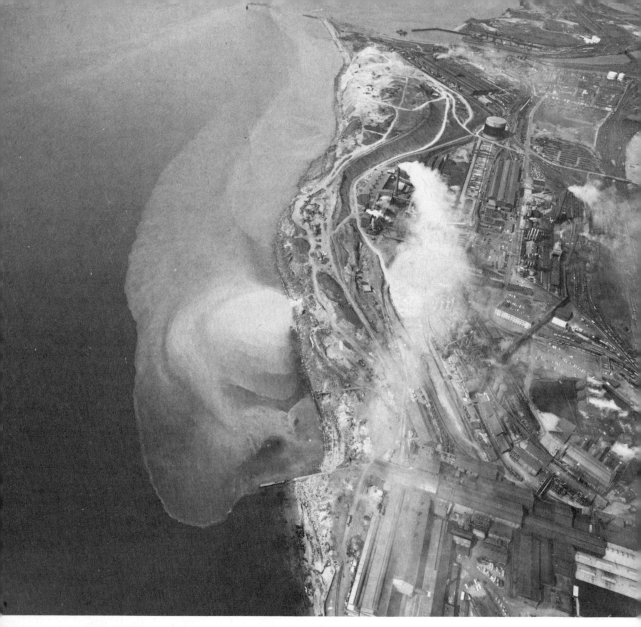

Figure 7.4 Industrial effluent   Discharge of effluent to the sea on the Cumbrian coast (Aerofilms).

posits of the Earth, modifying the natural distribution of elements in the ways already discussed (Fig. 7.4). A large part of this waste material, though unsightly, is not particularly harmful. Much organic waste is quickly rendered harmless by bacterial action and oxidation and can be recycled to maintain the fertility of cultivated land. The problems of waste disposal centre about three types of material: (i) very bulky waste such as the overburden removed during opencast min-

ing, (ii) waste in which toxic metals or other compounds are concentrated, and (iii) radioactive waste. The principal problems resulting from the large-scale generation of fragmental rock waste are those connected with the maintenance of stability in tip heaps and with the rehabilitation of dumps and stripped areas from which soil has been removed. These problems have been touched on in Chapter 6 and will not be discussed again.

118

Pollution results from the release by man of substances likely to affect human health or to damage ecological systems in general. The contamination of water by the discharge of untreated sewage into surface or groundwater systems is among the most serious forms of pollution, especially in urban societies, and is largely responsible for the spread of water-borne diseases such as dysentery, cholera and typhoid (Section 2.1). The construction of enclosed sewers diverting liquid waste directly to treatment plants was perhaps the most effective 19th-century public health measure in Britain. Today, less than half the population of the developing countries has access to a safe water supply and the construction of sewage systems is still urgently needed. Standard methods of treatment that involve bacterial decomposition and oxidation of organic matter generally allow the liquid end product to be returned safely to rivers and even to contribute to domestic water supplies. These methods, however, do not effectively remove synthetic compounds such as detergents which tend to pass on into the surface drainage. Solid matter consolidated in settling tanks is sometimes suitable for use as organic fertiliser, but when contaminated by heavy metals from industrial effluents it must be otherwise disposed of; much of the residue from treatment plants is dumped at sea.

Table 7.4 Metals in stream sediments: effects of mineralisation and mining.

	As	Cd	Co	Cu	Pb	Zn
mode for England and Wales (ppm)	7–15	0–1	10–20	15–30	0–40	50–200
Redruth area, Cornwall (ppm, maximum)	>150	>5	>80	>120	>320	>800

Surface and ground water may also be polluted by factory effluents not routed through the sewage system, by soluble fertilisers (especially nitrates) washed off farmlands and by metal-rich solutions leached out of mine dumps and industrial waste tips (Fig. 7.5; Table 2.1). Stream sediments down stream of old mining areas or industrial complexes record contamination by anomalous levels of toxic metals and/or radio-elements. The effects of industrial activity were illustrated in Figure 1.6; an example of contamination in a mining area is provided by stream sediment analyses in the old tin mining districts of Cornwall as shown in Table 7.4.

Atmospheric pollution resulting from the combustion of fossil fuels and the emission of fumes from factories, refineries and smelters leads to the temporary build-up of gases, droplets and solid particles (smog) in the air

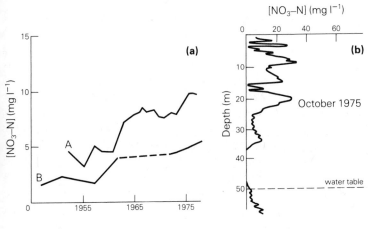

Figure 7.5 Contamination of ground water by nitrates from fertilisers (a) The increase of nitrate 1945–75 in a Triassic sandstone aquifer in Worcestershire (A) and the Chalk aquifer of Surrey (B). (b) Variations with depth in chalk at an arable site in Hampshire (based on Wilkinson, W. B. and L. A. Greene 1982. *Phil Trans R. Soc. Lond. B* **296**).

and, in due course, to the contamination of soil and water by materials returned to Earth with the rain. In the aftermath of the Industrial Revolution annual deposits of soot totalled about $1\,kg\,m^{-2}$ in regions such as London, a nuisance greatly reduced during the past half-century by legislation forbidding the use of coal in open fires. Up to $12\,g\,m^{-2}$ of sulphur are still deposited annually down wind of industrial areas from sulphur dioxide emitted to the atmosphere (sulphur averages about 1.5% and 3.0% respectively in coal and fuel oil). When oxidised to sulphate, SO_2 returns to Earth as 'acid rain' which promotes the deterioration of stonework (Fig. 5.1) and may damage vegetation and the faunas of rivers and lakes. Emissions from smelters and factories, especially those in use during the 19th and early 20th centuries, carry a variety of toxic metals which are reflected also in anomalies in soils and stream sediments down wind (Fig. 7.6). Such anomalies are found near zinc smelters in South Wales

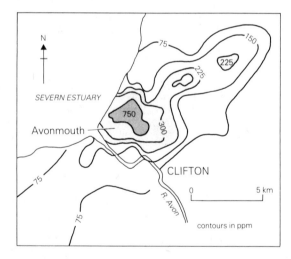

Figure 7.6 Effects of atmospheric pollution The pollution of soil, surface waters and vegetation resulting from the emission of industrial waste to the atmosphere. The figure shows variations in the cadmium content of sphagnum moss down wind of the industrial complex of Avonmouth: contours show relative values of Cd absorbed over a period in 1972 during which southwesterly winds prevailed (based on Little, P. and M. H. Martin 1974. *Environ. Pollut.* **6**).

and Avonmouth (Bristol Channel) and are recorded more dramatically by blackened rocks and stunted trees in the nickel mining area of Sudbury, Canada (Section 4.3.2).

The pollution of water and air by industrial processes is best controlled by the removal of toxic substances at source. The techniques by which liquid and gaseous effluents can be cleaned are, with a few exceptions, fairly well established, and in many countries maximum permitted levels for the emission of many substances are laid down by law. In Britain, practice is governed by the idea of the 'best practicable means' which aims to hold pollution at the levels that best serve the total interest of society. Elsewhere, legal limits may be set at very low levels which cannot always be achieved.

Radioactive waste from the fission process used in nuclear power stations (Section 3.4) poses special problems because it contains radioactive isotopes not normally present in our environment. On removal, the spent fuel has a high radioactivity owing to the presence of short-lived isotopes with half lives of up to a few years. The decay of these components leaves a much less radioactive residue – an initial radioactivity of 2 million curies in fuel removed from a Canadian reactor is reduced to 1600 curies within ten years – incorporating longer-lived isotopes such as iodine 129, caesium 135 and neptunium 237 with half lives of up to a million years.

During the initial phase of high radioactivity, waste is stored above ground in sealed containers resistant to corrosion. The need to ensure that the residue remains isolated from the biosphere for periods of at least a million years has led to discussion of the suitability of geological sites as repositories. To ensure that the stored waste does not become involved in normal geological processes, reactions within the container or between the container and the adjacent rocks or pore waters must be minimised. Dispersal of radionuclides in a glass or (geologically a more attractive proposal) in synthetic minerals of stable species, and packing in containers of inert metals or corundum are among the possibilities under consideration.

Long-term storage requires a geological setting from which any fluids that did become contaminated in the event of leakage would find their way to the surface only slowly and in a much diluted form. This *desideratum* has evoked ideas of disposal in a Benioff zone, where a descending crustal plate could remove the waste from the surface, or near the down-current of a submarine hydrothermal convection cell. Other considerations point to the selection of tectonically stable sites (aseismic, low geothermal gradient) and to the need for the enclosing rocks to be either strong and unfractured (massive granite) or impermeable and self-sealing (clay, evaporite). Burial at depths of more than 1 km would reduce the possibility of leakage because permeability decreases with depth.

The long timespan necessary for the isolation of radioactive waste and the unfamiliar nature of the elements concerned have given rise to widespread public anxiety. The need to dispose of spent fuel already in existence emphasises the need for continued research despite the unpopularity of the subject.

References

Bridges, E. M. 1970. *World soils*. Cambridge: Cambridge University Press.

Detwyler, T. R. 1971. *Man's impact on environment*. New York: McGraw-Hill.

Howard, A. D. and I. Remson 1978. *Geology in environmental planning*. New York: McGraw-Hill.

Hudson, N. 1971. *Soil conservation*. London: Batsford.

Tank, R. W. (ed.) 1973. *Focus on environmental geology*. Oxford: Oxford University Press.

Thornton, I. (ed.) 1982. *Applied environmental geochemistry*. New York: Academic Press.

8 Methods of exploration and site investigation

The effective application of geological knowledge to the practical matters discussed in Chapters 2–7 of this book depends largely on the availability of techniques capable, first of locating specific targets such as ore deposits; secondly, of building up an adequate knowledge of sites at which mining or construction are taking place; and, thirdly, of assessing potential or actual geological hazards. Reconnaissance surveys which provide data from large areas quickly and cheaply are used to narrow the field of investigation in the early stages of any exploration. The detailed investigation of individual sites, once identified, is designed to yield facts and figures comprehensive enough to provide a basis for further development. Most reconnaissance methods and several techniques used during the preliminary stages of site investigation have applications in many different fields and are dealt with together in this chapter (Table 8.1). More specialised methods, such as are used in the water and civil engineering industry, have been mentioned in the appropriate places and are not referred to again here. The notes that follow are intended to give the reader a general idea of how the common methods of investigation work. They do not cover the mathematical background or the practical details needed to operate these methods, for which more specialised books are available (see References).

Almost all successful applications of the Earth sciences have to take account of the regional and/or local geological setting – the sum total of factors determining the distribution and structure of the rocks concerned. All the facts that can be obtained from published maps, reports and papers ('desk study') or from direct observation in the field and the laboratory may be relevant to the geological setting. The methods of enquiry are much the same whether or not the investigation has an 'applied' slant and as they are dealt with in many textbooks on Earth science (see References for Ch. 1) they are not considered in detail here. The techniques used more specifically in exploration and site investigation depend on variations in the properties of rock materials that serve to identify relevant deposits or structures. Table 8.1, which classifies techniques according to the properties selected for investigation, provides the basis for the treatment adopted below.

8.1 Geophysical methods

Techniques that depend on measuring the physical properties of rocks have the advantage of yielding information about rocks and structures concealed below the surface which are not accessible to direct observation. They are widely used in the oil industry and are also important in mineral exploration and in the preliminary investigation of sites for construction and mining. Geophysical methods of recording changes going on below the surface are also widely used to monitor

Table 8.1 Exploration methods: summary.

Method	Principal applications
Geophysical methods	
seismic surveys (reflection and refraction)	elucidation of subsurface structure, especially: structure of sedimentary basins, recognition of key horizons, unconformities, folds, faults in oil and gas fields: exploration of superficial deposits of construction sites
seismicity records	monitoring of active volcanic centres, fault zones
gravity	elucidation of regional structure in sedimentary, igneous and metamorphic terrains, useful at reconnaissance stage of exploration for minerals, hydrocarbons: identification of anomalies related to buried igneous centres, ore deposits, salt domes (oil traps)
magnetic	elucidation of regional structure in igneous and metamorphic terrains, useful for reconnaissance for mineral deposits; identification of anomalies related to buried igneous centres, iron formations
electrical and electromagnetic	resistivity surveys, principally to locate ore bodies and in borehole logging
	self potential (SP) and induced potential (IP) surveys to locate ore bodies
	electromagnetic surveys to locate ore bodies and detect rise of magma in volcanic centres
radiometric	prospecting for uranium, thorium; borehole logging
	investigation of areas with anomalous radioactivity
Remote sensing	airborne geophysical reconnaissance: regional structure
	topographical surveys
	geological reconnaissance
	surveillance of potential hazards such as volcanic centres
	monitoring environmental changes
Subsurface sampling boreholes	elucidation of regional succession and structure; mud logging for general lithology, microfauna (oilfields) logging of core for structural and petrogenetic detail (oilfields, mineral exploration, site investigation)
augering	investigation of weathered mantle, superficial deposits
Geochemical and mineralogical methods reconnaissance surveys	stream- or lake-sediment samples, water samples, reconnaissance for ore deposits, basis for investigation of geochemistry in relation to plant, animal or human health
	heavy mineral concetrates
local surveys	location of mineral deposits, investigation of possible geochemical hazards
periodic sampling	quality control (water and effluents): effects of pollution

potentially hazardous developments, both natural and man-made.

8.1.1 Seismic surveys

Shock waves generated by earthquakes or by man-made explosions travel through rock at rates determined by the seismic velocity, usually expressed in metres or kilometres per second. The seismic waves are reflected or refracted at the junctions of rock units with differing seismic velocities and a proportion of the waves so deflected return to the surface where they are received by appropriate sensors (geophones). Information about the velocities of the rocks traversed and the depth and inclination of the interfaces between lithological units is provided by the record of the travel times of impulses on the return journey from shotpoint to one or more geophones (Fig. 8.1). A seismic profile is built up from observations made at a number of stations. Most seismic surveys are concerned with the faster (longitudinal) P waves rather than the S (shear) waves emitted along with them.

Seismic ground surveys involve the use of explosives or mechanical thumpers at a number of sites and the deployment of geophone arrays to record returns from each shot. For the shallow penetration needed, for example, to determine the thickness of alluvium or till over bedrock in the foundations for a proposed bridge or other construction, the impulse needed can be generated with a sledgehammer and the distances between sites may be no more than a few metres. The deep penetration needed to establish the structure

of an oilfield reservoir to depths of many kilometres, on the other hand, calls for heavy equipment cumbersome to move and unsuited for use in populous areas. Land surveys therefore tend to be expensive: the cost of a 100 km profile in Britain around 1980 was several hundred thousand pounds. Marine surveys in which shots are fired at sea and a geophone array is towed by the survey ship are notably cheaper.

Seismic velocities have characteristic ranges in rocks of different types (Fig. 8.1). Seismic methods are useful in distinguishing between unconsolidated surface materials and bedrock, between sedimentary and metamorphic or igneous rocks and, in some circumstances, between more and less thoroughly lithified sedimentary rocks. The high velocity in water relative to that in air is reflected in velocity contrasts between saturated and unsaturated rocks, where porosities are high. The principal uses of the methods are in oilfield exploration (see Section 3.4) and in engineering geology. **Reflection surveys** which depend on 'bouncing' waves off lithological boundaries at which there is a marked velocity contrast (Fig. 8.1) provide a powerful tool for the exploration of sedimentary basins. A seismic profile, when processed by appropriate mathematical methods, provides data on the thickness of units with characteristic velocities, the presence or absence of distinctive reflectors (which may be identified in stratigraphical terms from borehole evidence), the positions of disconformities or unconformities and the effects of folds and faults. The cross section of North Sea oilfields and the schematic illustration of seismic stratigraphy shown in Figures 3.13 and 3.9 are derived mainly from reflection seismic data. **Refraction seismic methods**, which depend on the study of waves refracted along lithological boundaries (Fig. 8.1) are of special use in defining oilfield traps such as anticlines and salt domes.

Observations of natural seismicity in and

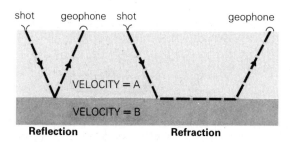

Representative velocities of P (longitudinal) waves (m s^{-1})

air	330
water	1450
soil, sand	170–800
shale, sandstone, limestone	1500–4000
metamorphic rocks	>5000
granites	>5000

Figure 8.1 Seismic waves Simplified paths of seismic waves used in the reflection and refraction methods of seismic surveying.

around active fault zones or volcanoes depend for their value in forecasting hazards on the continuous recording of minor tremors over long periods of time. The seismometer used for this purpose (like the instruments used to determine deep structure) magnifies impulses transmitted from the ground by means of arms pivoted on a stable base and records the resulting seismogram on a revolving drum. Changes in the magnitude, spacing and location of minute tremors relative to the known background may give warning of an impending earthquake or eruption.

8.1.2 Gravity surveys

The value of gravity g obtained at a point on the Earth's surface depends not only on factors such as latitude, altitude, topography and regional structure but also on the relative densities of the underlying rocks. A gravimeter survey covering many sites provides the data from which **gravity anomalies** related to the underlying rocks can be recognised when the appropriate corrections for other variables have been made. Bodies of relatively dense rocks give positive gravity anomalies ('highs') and relatively light rocks give negative anomalies ('lows'); the magnitude of local anomalies ranges up to a hundred or so milligals. Ground surveys at closely spaced sites are useful for detecting anomalies originating not far below the surface. Marine, and especially airborne, surveys also detect anomalies originating at depths of at least 10 km. The results are commonly summarised on contour maps, after correction for extraneous variables (the free air, terrain and Bouguer corrections give notional values for g at sea level and remove effects due to topographical irregularities). Interpretation depends on knowledge of the relative densities of the rock present, obtained either by reference to average values (Fig. 8.2) or by the direct measurement of samples. By assigning suitable densities, the size and shape of a rock mass needed to create an observed anomaly can be modelled mathematically (Fig. 8.2). Repeated gravity surveys over active volcanic centres may detect changes due to the rise of magma prior to an eruption.

The gravity method is applicable to structural studies in both sedimentary basins and metamorphic or igneous terrains. Regional anomalies may help to define the configuration of a basin and the sites of basement ridges near which hydrocarbons or mineral deposits may be found; for example a regional high on the continental shelf west of the Shetland Islands defines a basement ridge separating two potentially oil-bearing basins. Gravity surveys are, naturally, used most often in the search for targets known to be significantly denser or lighter than the enclosing rocks. The high relative densities of most ore minerals (Table A1.1) cause many ore deposits to give strong positive anomalies. The small size of most ore bodies limits this direct approach to the exploration of known orefields where readings can be spaced at intervals of a few metres or tens of metres. Igneous intrusions associated with mineral deposits provide larger targets; for example, many basic intrusives give positive anomalies and some granites give negative anomalies (cf. Fig. 8.2). Salt domes (which may provide traps for hydrocarbons) commonly give pronounced negative anomalies.

8.1.3 Magnetic surveys

Magnetite (Fe_3O_4) and other magnetic minerals impart to their host rocks a magnetism that varies according to composition, age, origin and texture. Rock magnetism has both magnitude and direction, the latter being determined by position relative to the past and present magnetic poles of the Earth. Positive and negative **magnetic anomalies** (expressed in gammas and ranging up to more than ± 1000) record the magnetism of rocks in which the alignment of magnetic particles shows opposite orientations.

A simple compass needle which is deflected in the vicinity of magnetic rocks

125

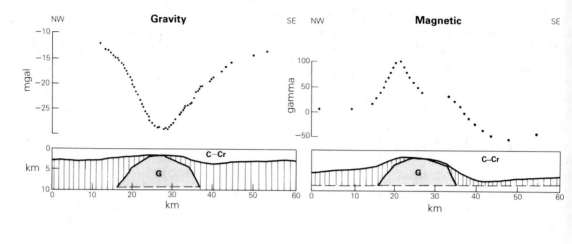

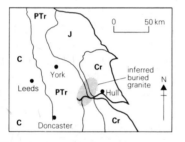

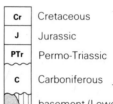

		cover – relative density 2.25–2.6, non-magnetic
Cr	Cretaceous	
J	Jurassic	
PTr	Permo-Triassic	
C	Carboniferous	

basement (Lower Palaeozoic and Precambrian)– relative density ~ 2.75, magnetic
G = inferred granite – relative density 2.6, magnetic

Figure 8.2   Use of gravity and magnetic surveys   The negative gravity anomaly and complex magnetic anomaly defined by readings along a NW–SE profile in the Market Weighton area are interpreted in terms of a granite beneath the unconformable cover of Carboniferous to Cretaceous strata. The schematic shapes of the inferred granite are those corresponding to the calculated curves which best fit the observed anomalies (based on Bott, M. H. P., J. Robinson and A. Kohnstamm 1978. *J. Geol Soc. Lond.* **135**, 535–44).

Representative relative densities

water	1
sand	2
lithified sedimentary rocks	2.2–2.8
halite, gypsum	2.2–2.3
metamorphic rocks	2.6–3.0
granite	~2.6
basic and ultrabasic rocks	2.8–3.3
metallic sulphides	>4.0 (Table A1.1)

was used in Sweden as a prospecting tool for iron ore as early as the 17th century. Modern *magnetometers* in which the compass needle is replaced by one of a variety of magnetic devices, are sensitive to very small variations in magnetism. Their principal economic uses are in metamorphic and igneous terrains where aeromagnetic surveys provide remarkably detailed information concerning the trends of marker horizons, the positions of faults and the distribution of igneous intru-

sions. Marine surveys have played a crucial role in establishing the structure of the oceanic crust, and particularly in locating spreading centres and transform faults. All these data help to identify sites favourable to the occurrence of metalliferous mineral deposits, while detailed ground surveys may locate iron ores and certain other deposits directly. Observations (after the application of terrain corrections analogous to those used in the processing of gravity data) are summaised on contour maps and can be used to model the forms of rock units responsible for individual anomalies as illustrated in Figure 8.2. A major asset of magnetic methods arises from their ability to 'see through' a cover of non-magnetic or weakly magnetic sedimentary rocks to reveal structures in the underlying basement. For example, the Precambrian Krivoi Rog banded iron formations, which are exposed in the Ukraine, are associated with a 1000 km linear magnetic anomaly that has enabled them to be located beneath the Phanerozoic cover of the Russian platform.

8.1.4 Electrical and electromagnetic methods

The capacity of rock to transmit electrical currents and to sustain natural or induced differences of electrical potential provides the basis for a variety of exploration methods. **Resistivity surveys** involve passing a current from a generator into the ground via two electrodes and measuring the voltage reduction at a second pair of electrodes inserted between them. Lateral and vertical variations are investigated by shifting the electrode pairs laterally or by increasing the spacing between pairs. Induced polarisation (IP) surveys make use of the fact that voltage differences produced by this mechanism do not immediately die away when the current is switched off, as a result of the build-up of charges on grains of good conductors.

Resistivity surveys have two principal applications in mineral exploration and engineering geology, which depend on the low resistivity (or in other words, the high conductivity) of metallic ore minerals (with the exception of sphalerite ZnS) relative to silicates and of saturated rocks relative to dry rocks. In mineral exploration, resistivity and IP surveys of ore bodies may be combined with **electromagnetic surveys** in which an instrument similar to the metal detectors used in security checks is employed. Electromagnetic waves generated by the instrument induce currents in conductors such as sulphides which are recorded by sensors in the detector.

Resistivity surveys are used in engineering geology to investigate unconsolidated materials with high porosity where water (especially salty water) acts as a relatively good conductor. Downhole logging in oilfield wells commonly includes resistivity surveys designed to identify porous horizons, especially those containing brine.

8.1.5 Radiometric surveys

The radioactive breakdown of such naturally occurring radioelements as uranium, thorium, potassium and radon, and of the radioisotopes produced in nuclear reactors, leads to the emission of sub-atomic particles which can be detected by appropriate sensors. A simple **spectrometer** measures total radioactivity at a given site. Multichannel instruments can discriminate between emissions from different sources and are used both as prospecting tools for uranium (usually detected indirectly by emissions from the daughter element bismuth 214) and for the purpose of investigating potentially harmful sources of radiation.

Airborne radiometric surveys detect regional anomalies in uranium distribution which have proved useful in the Canadian Shield in identifying areas suitable for more detailed exploration. Ground surveys may locate individual uranium deposits exposed at the surface and, by using different channels, can be used to investigate a number of environmental problems, notably those connected with the dispersal of radioactive materials from spoil heaps or waste dumps and the monitoring of radioactivity around reactors and waste storage sites. The detection of radon (gas) escapes from fault zones is potentially useful in the prediction of earthquakes.

8.2 Remote sensing and photointerpretation

Aerial surveys of the Earth from aircraft and orbiting satellites provide unrivalled sources of information about surface features of the lithosphere and its vegetation, and about the hydrosphere and atmosphere. Many of these surveys have geological applications and a number of satellites (Landsat, Geosat, etc.) have been used principally to collect information on natural resources. The techniques of surveying, transmitting and mathematical treatment used in remote sensing have been developed largely by the defence, space and communications industries, using instruments mounted in satellites or aircraft or towed behind aircraft in a contraption known as a 'bird'.

The geological advantages of these methods are of three kinds. They provide information on the topography and geology of little-known terrains which can be used to increase the effectiveness and decrease the cost of reconnaissance on the ground; they reveal structures that may be invisible at ground level; and they investigate properties not easily made use of in ground surveys.

Aerial gravity, magnetic and radiometric surveys have been mentioned in previous sections. The methods remaining to be dealt with are those concerned with the observation of rays with a wide range of wavelengths that are reflected by the Earth from the Sun or from sources carried by the satellite itself. A single satellite image picturing a part of the Earth's surface may cover an area over 100 × 100 km^2; an air photo taken from a height of a few kilometres covers an area of a few square kilometres. Data referring to different **spectral bands** (i.e. from differing wavelengths) can be selected, enhanced and recombined by computer manipulation.

Black and white or colour photographs produced by rays within the spectrum of visible light are widely used both for carto-graphic purposes and in geological reconnaissance. Stereoscopic cover from which a three-dimensional image is produced when images of adjacent overlapping prints are optically superimposed by means of a stereoscope, provides a vivid and realistic, though distorted view of surface features. Where distortion has been reduced by the application of appropriate corrections for variations in height and topography, the cover provides a basis from which topographical maps can be made. Where vegetation is sparse, many rock units can be tentatively outlined by reference to variations in colour and surface 'texture' (related to style of weathering and fracturing) and structural observations such as the dip of strata and the traces of folds and faults can be inferred from the small topographical variations resulting from differential erosion (Fig. 8.3). Even in heavily vegetated or cultivated regions, the broad outline of the structure may be indicated by topographical features and variations of rock type by variations in the flora. Photointerpretation based on observations of these kinds is widely used to plan and focus the work of field survey teams during a reconnaissance programme, the observations being checked later against 'ground truth'.

Spectral bands outside the visible light range reveal features not normally accessible to the human eye. Multispectral images and false colour images generated by assigning an arbitrary colour to one or more spectral bands are especially sensitive to variations in the nature and wellbeing of the plant cover and are used for a variety of ecological, agricultural and environmental purposes. Of particular interest from the geological viewpoint are radar imagery which is not obstructed by a cloud cover, and infrared imagery, which detects natural and man-made heat sources. The infrared method has been used to study the dispersion of fumes emitted to the atmosphere and of (possibly contaminated) effluents discharged into lakes or estuaries.

Figure 8.3 Photogeological methods This vertical air photograph of a part of north-west Scotland covers an area of several square kilometres. To the south of Lochan Fada, subhorizontal Torridonian sandstone, resting on a metamorphic basement, is identified by terraced features. A NNE fault separates the Torridonian outcrop from Cambrian strata which are overthrust by basement gneisses (reproduced from an Ordnance Survey aerial photograph, Crown Copyright reserved).

The value of remote sensing methods is greatly increased by the opportunities they provide to record natural or man-made changes over a period of weeks, months or years by comparing images recorded on successive orbits. Many sudden events such as earthquakes, volcanic eruptions or landslides are heralded by detectable slower changes as, for example, an eruption may be preceded by the rise of magma bodies with consequent increase of heat emission visible to infrared sensors. The effects of disasters may be most rapidly assessed from the air and, finally, progressive slow developments such as the migration of sandbanks or the advance of desert conditions may be accurately recorded.

8.3 Sampling of subsurface materials

Although structural analysis and geophysical surveys allow one to predict the distribution of rocks at depth, direct sampling of subsurface material provides the only sure means of checking the predictions based on these methods. The sinking of deep boreholes or wells – by far the most important method of subsurface sampling – is extremely costly (a 7 km hole in Britain cost around £6 million in 1980) and is used principally in association with seismic profiling in the oil industry. Boreholes of a few hundred metres depth are used to investigate favourable anomalies in the search for ore deposits and augers (essentially, large corkscrews which retain samples on their whorls when withdrawn from the ground) for the exploration of weathered or unconsolidated material at construction sites. Wells used for the transmission of water or hydrocarbons and for the recovery of geothermal heat are usually cased in by cementing material or by tubes fitted into the hole and are controlled by pumps. Most exploratory boreholes are sealed off after completion.

Drilling is carried out by means of a 'bit' attached to an extensible rod or line and suspended from a stable rig. The bit, hardened by diamond chips (diamond drilling) or made of special steel, is rotated as it is driven into the hole, its passage being lubricated by water (for short holes) or by mud improved by various additives such as baryte. The solid rock in the hole is detached either in the form of a continuous core that is withdrawn intact at intervals or in the form of loose chips brought to the surface along with the drilling mud. In mining and mineral exploration much of the core is recovered intact. Diamond drilling is the principal method used and the cores are generally only 2–6 cm in diameter. In oilfields, most holes are broader, much of the rock material is recovered as chips, and continuous coring is reserved for critical sections such as those traversing a known seismic reflector.

A continuous record is kept of the progress of drilling on which the characteristics of the rocks traversed and the depths to lithological boundaries are entered. Mud logging of fragmental material is carried out as drilling takes place. Detailed studies of the core add data concerning such features as sedimentary structures and dip in sedimentary basins, intrusive contacts, cleavage and hydrothermal veins in mineralised terrains. These initial studies may be followed by the extraction and identification of microfossils and any macrofossils obtained, by analysis of selected samples, by isotopic dating, and so on. Geophysical logging (downhole logging by resistivity, radiometric and other methods) is commonly carried out during drilling in sedimentary basins (cf. Fig. 3.10, Table 3.3). An essential aspect of all borehole logging is that of determining the orientation of the hole. Most holes are intended to be vertical or, in some mineral exploration work, oblique but rectilinear. In practice, many diverge from their intended course by as much as 90° and their actual course must be known before estimates of thickness and dip can be confirmed.

8.4 Geochemical and mineralogical methods

Variations in the abundances of major and trace elements in rock, soil and other materials at the Earth's surface have a direct bearing on the fertility of the land and also offer clues to the whereabouts of metalliferous and non-metalliferous deposits. Geochemical exploration methods which exploit these relationships are considered here along with a few methods that depend on the distribution of key minerals. They are used principally for prospecting and in relation to problems of agriculture, medicine and pollution.

Reconnaissance geochemical surveys which cover large areas depend on the collection of samples which represent a varied terrain in a statistically valid way. Sampling of the bedrock at grid intersections or according to other predetermined plans seldom proves satisfactory in metamorphic and igneous terrains where small-scale variations are common. Stream and lake sediments derived from the erosion of exposed rocks and surface deposits on the other hand, represent naturally formed samples of the materials of the drainage basin and surveys based on the analysis of stream sediments provide a good general picture of the distribution of many elements at sample densities of 1–2 sites per km². They are often combined with sampling of stream waters, which carry dissolved matter from the drainage basin together with organic solutes, and in some instances, by sampling of vegetation (biogeochemical surveys). Sampling sites are generally chosen above the confluence of tributaries in order to simplify identification of the sources of anomalies (see below). Heavy mineral concentrates obtained by the time-honoured prospector's method of **pan-** ning (that is, by swirling sediment in a shallow dish in a water current strong enough to wash out the dominant light minerals such as quartz) provide additional evidence as to the nature of the rocks upstream. The very large numbers of samples acquired (50 000 were collected for a reconnaissance survey of England and Wales) are analysed by rapid automatic methods tested for reliability at several stages. Data for up to about 35 elements are handled automatically and can be presented in maps or computer printout. Geochemical anomalies identified at the reconnaissance stage become the targets for more detailed investigation.

The elements most commonly dealt with in geochemical surveys include the metals and other elements that may be targets for prospecting (e.g. Cr, Cu, Fe, Ni, Pb, Sn, U and Zn), the elements commonly associated with mineral deposits which may be regarded as **pathfinders** (e.g. arsenic gives anomalies near many sulphide deposits), and associated elements that provide clues to the origins of important rocks in a mineralised area. Lastly, elements known to be essential to nutrition or harmful to health, including lead, cadmium and other heavy metals as well as radioisotopes originating in nuclear reactors, are recorded where the objects of the survey are mainly environmental.

Although many of the elements in stream sediments reflect more or less faithfully the compositions of the bedrock from which the sediments are derived, allowance has to be made in interpreting geochemical data for distortions resulting from surface processes. Iron and manganese, for example, tend to concentrate in stream sediments where the water is acid and organic matter abundant. Uranium often forms soluble compounds in oxidising conditions and, with zinc, fluorine and a number of other elements, therefore tends to be under-represented in stream sediment. Complementary analyses of stream waters and direct examination of the sample may allow such distortions to be recognised.

Concentrations of metals or uranium that appear, after consideration of the factors outlined above, to relate to the bedrock may be investigated by closely spaced sampling of stream sediments and waters designed to trace anomalies up the drainage system to their source (Fig. 8.4). Rock fragments in the stream bed, weathered mantle or glacial drift and heavy mineral concentrates from the stream sediment may reveal ore or gangue minerals or characteristic associates of the deposits being sought. The magnesian garnet pyrope, for example, is common in kimberlites but rare in most other rocks. It is many times more abundant than diamond and it forms an excellent marker for diamond pros-

pectors. The last stages of investigation may involve trenching and pitting and radiometric or other geophysical surveys.

Where the investigation concerns the environmental implications of geochemical anomalies, attention is focused on soil, water dust and waste accumulations rather than on bedrock, as it is mainly from these surface materials that elements are taken up by plants and animals. Routine sampling and analysis are carried out to check on the quality of water supplies and sewage or other effluents and similar techniques are used to monitor changes resulting from new development such as the opening of a municipal dump. Analysis for these purposes records pH, Eh

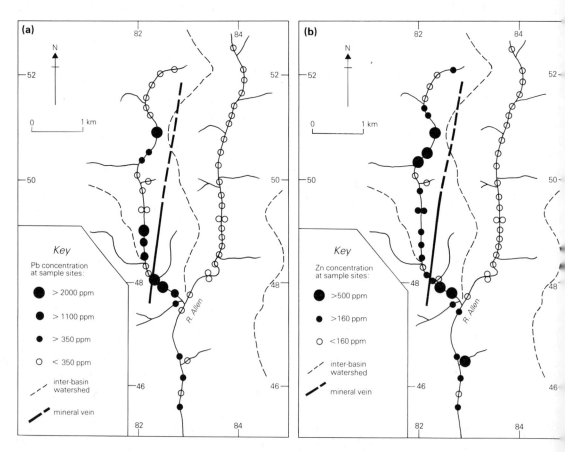

Figure 8.4 Stream sediment analysis   The sampling of the River Allen and its tributaries (north of Truro, south-west England) during prospecting reveals high levels of (a) lead and (b) zinc near a galena-bearing vein (M. Hale, unpublished data).

and a number of organic solutes, as well as toxic metals and bacteria. Long-term experiments involving the comparison of crops or stock raised in geochemically anomalous areas with controls elsewhere, together with investigations of the role of metals and other elements in the metabolic processes, call for methods that are beyond the range of the geological sciences.

References

Clark, I. 1979. *Practical geostatistics*. London: Applied Science Publishers.

Reedman, J. H. 1979. *Techniques in mineral exploration*. London: Applied Science Publishers.

Townshend, J. R. G. 1981. *Terrain analysis and remote sensing*. London: George Allen & Unwin.

9 Afterword: man in a geological context

The preparation of this book has given the writer a new appreciation of man as a geological agent. In this context, the antithesis often perceived between 'natural' and man-made happenings seems unreal. Man is, at present, the dominant species of the biosphere and the advent of industrial human societies has brought changes analogous to those initiated in Palaeozoic times by the evolution of rock-forming metazoans or in the

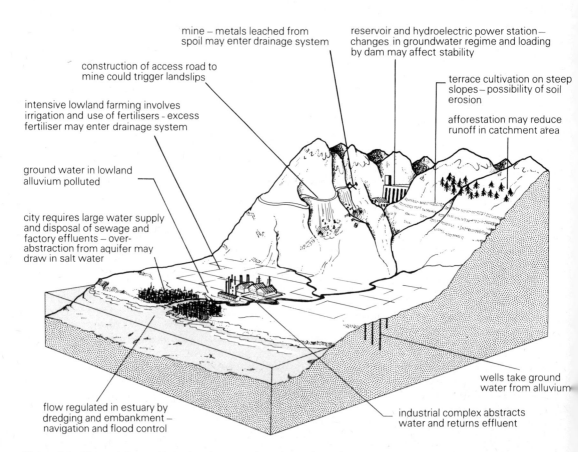

mine – metals leached from spoil may enter drainage system

reservoir and hydroelectric power station – changes in groundwater regime and loading by dam may affect stability

construction of access road to mine could trigger landslips

terrace cultivation on steep slopes – possibility of soil erosion

intensive lowland farming involves irrigation and use of fertilisers - excess fertiliser may enter drainage system

afforestation may reduce runoff in catchment area

ground water in lowland alluvium polluted

city requires large water supply and disposal of sewage and factory effluents – over-abstraction from aquifer may draw in salt water

wells take ground water from alluvium

flow regulated in estuary by dredging and embankment – navigation and flood control

industrial complex abstracts water and returns effluent

Figure 9.1   River regimes   Natural and man-made regimes along the course of a river: a schematic diagram.

later Palaeozoic times by the emergence of advanced land floras. Whether we like it or not, man is a geological agent and as such interacts with other agencies working on the land and (to a lesser extent) in the sea. The human influence on geological processes does, however, have three special characteristics: it extends into almost every surface environment; it operates on a short timescale; and it is, in principle, subject to rational control.

The extent to which man's need to find raw materials and to adapt the landscape to his requirements impinges on other geological processes may not have been made fully apparent in the previous chapters, where individual resources and activities were treated separately. The reader can restore the balance by mentally cataloguing all the human activities with geological repercussions which might be going on in a natural region such as a delta or coastal plain. Figure 9.1 illustrates the point with respect to an imaginary river system. It will be evident that some mechanism for planning and coordination of projects undertaken for different reasons will be needed if incompatible developments are to be avoided and adverse side effects minimised. Geological expertise is required at three stages in this process: first, to supply basic information in the form of geological, geophysical and geochemical maps and reports; secondly, to find and evaluate deposits of fossil fuels, metals, construction materials and other resources and to identify suitable sites for constructions such as dams; thirdly, to assess reserves, to forecast the long-term geological effects of existing or new developments and to recognise potential geological hazards. The provision of basic information is generally the responsibility of national governments, though major contributions may also be made by international agencies (especially in marine geology) by academic research workers and by large commercial organisations. Exploration and site investigation are handled mainly by govern-

ment agencies (especially in the Eastern bloc) and by international or national oil and mining companies and civil engineering firms. The final stage of assessment and forecasting involves political, social and economic, as well as geological, considerations. The weight given to each contributing factor varies according to the structure of the organisation concerned, the political and economic climate and the personalities involved; but the committees in or out of government which have to take decisions

Table 9.1 Geological and human timescales.

Minimum duration	Event (italic indicates human influence involved)
	meteorite impact
	earthquake, landslide, mudflow, *mine collapse*
	flash flood, *dam burst*
	storm, turbidity flow
	explosive volcanic eruption, pyroclastic flow
	rock burst, mine explosion
1 day	volcanic eruption
	soil erosion, hill creep
	pollution of lake or river
1 decade	migration of sandbars and sandbanks
	silting of harbours
	depletion of aquifer, entry of salt water
	decay of short-lived radioisotopes
1 century	sea-floor spreading >10 m
	rise or fall of sea level
	climatic change
	accumulation of peat, alluvium, beach deposits
	decay of long-lived radioisotopes
1 million years	sea-floor spreading >10 km
	new species evolve
	maturation of oil and coal
	thick sedimentary successions
	large volcanoes built
	ore-forming processes?

Figure 9.2   A braided river in Iraq   The photograph illustrates the effect of repeated migration of channels which may encroach on settlements and cultivated land (Aerofilms).

on questions with a geological dimension seldom include Earth scientists among their number. In these circumstances it is especially important that non-geologists among both the decision makers and the general public should have some familiarity with geological concepts.

A common obstacle to the appreciation of geological realities is provided by the slowness of geological processes relative to the human lifespan. The idea of the natural scene as something unchanging ('as old as the hills ...') is so deeply rooted in the human mind that the inherent instability of most surface environments is often lost to view. In reality, many geological events are completed within periods measurable in years, or even days, the slow average rate of change being accounted for by the occurrence of long intervals between such events. Table 9.1, which compares the timespan of natural events with those triggered by human intervention,

(a)

(b)

Figure 9.3 Rehabilitation (a) The in-filling of a quarry with industrial waste. (b) The conversion of a flooded pit for recreation (Crown Copyright reserved).

shows that catastrophic earthquakes, floods and volcanic eruptions occur at rates comparable with those of dam bursts, rock bursts and mine collapse. Less dramatic developments connected with the rise or fall of sea level or the deposition of sand bars produce significant changes over periods of a few centuries, well within the range of human interest (Fig 9.2). The traditional image of a stable Earth must be set aside when long-term developments are under consideration.

A second source of misapprehension arises from the undoubted fact that many of the conventional sources of fuels and metals obtained from the Earth appear likely to run out within a comparatively few generations. In principle, most rocks and other resources are renewable because the processes responsible for their formation continue to operate; but, as Table 9.1 suggests, the natural rates of replacement may be so much slower than rates of consumption as to render them, for practical purposes, non-renewable. The more rapidly recycled resources such as water and soil nutrients appear to remain roughly in balance on a global scale, although severe shortages are recorded at particular times and places. Bulk construction materials, though available in vast quantities, are becoming scarce in many parts of the developed countries where demand is greatest. The conventional sources of fuels and of many metals are expected by many people to be exhausted within the foreseeable future. While it is probable that supplies from these sources will continue to be forthcoming for longer than the more pessimistic forecasts suggest (as a result of new discoveries and improved methods of extraction) the gap between rates of renewal and rates of consumption is unlikely to be closed. The total amount of metals extracted during the entire course of human history is, however, insignificant when compared with the amounts of these metals contained in the upper parts of the Earth's crust and the sea. Thus, for example, the estimated 100 000 tonnes of gold produced over five centuries since the discovery of America must be set against the 10 000 *million* tonnes of gold contained in the sea. Figures such as these suggest that so far as metals are concerned supplies from what are now unconventional sources may become of crucial importance in the future. In the present state of technological knowledge, the recovery of metals from very low-grade ores, and especially from silicate minerals, is prohibitively expensive in terms of energy requirements; but if means could be found to extract them from these sources, metals would no longer be in short supply. As regards fuels, the unconventional hydrocarbon deposits in oil shales and oil- or tar-impregnated sands (from which oil cannot be extracted by the customary methods of pumping) provide huge reserves which have only recently begun to be exploited on a significant scale. The radioelement uranium, the basis of nuclear power schemes, is, like the metals mentioned above, widely distributed in common rocks from which it may ultimately prove to be recoverable (Table 1.3). In the geological context, the resources of the Earth do not seem to be close to exhaustion, though means to exploit some of them have yet to be found.

Strategies intended to make the best use of

Table 9.2 The National Coal Board's ladder of exploration, 1981. Successful prospects move up the ladder from stage 1 to stage 6 (condensed from Commission on Energy and the Environment 1981. *Coal and the environment.* London: HMSO).

stage 1: potential prospects	coal is believed to exist at or below a particular site (many areas)
stage 2: preliminary exploration	coal is known to exist, but its extent is unknown (north of Carlisle, Firth of Forth, north-west of York, north-east of Witney, Oxfordshire, etc.)
stage 3: intensive exploration	boundaries of coal are known, results justify further exploration (offshore Northumberland, East Staffordshire, Gainsborough, Lincs, Banbury, Oxfordshire, East Nottinghamshire)
stage 4: feasibility study	mining engineering and economic assessment, identification of environmental impact (South Warwickshire, East Yorkshire, Kincardine)
stage 5: planning	application, with government support, for outline planning permission, based on specific proposals for mining (Belvoir, north-east Leicestershire – minimum time-period to full production from first mines estimated at 8 years)
stage 6: development	outline planning permission granted, preparations for production (Selby, Yorkshire – stage 5 began in 1974, full production expected in late 1980s)

Figure 9.4 Climatic change in historic times This ruined city in Cyrenaica recalls the time when North Africa was a major grain-producing region for the Roman Empire (Aerofilms).

the Earth's resources during the next few centuries must be designed to work with, rather than against, the geological realities. The practical procedures for implementing a decision to begin on a major new development are, in a modern society, complex and long-drawn out. Table 9.2 shows, for example, that more than a decade separates the first and last stages in the development of a new British coalfield. The considerations outlined above suggest that, on the one hand, there should be a gradual shift towards the discovery and assessment of resources of new kinds and, on the other, a more co-ordinated effort to maintain or restore the stability of geological environments that have been disturbed by human activities (Fig 9.3). To control or modify a geological process, however, one

139

must first understand it; and geologists are working close to the frontiers of knowledge in many fields of practical importance. Too little is known, to take random examples from among present problems, about the reactions likely to follow the recharge of an aquifer with water differing in composition from the original pore fluid; about the extent and variation of deep-sea metalliferous sediments; about the relationships between soil chemistry and human or animal health; and about the behaviour of radioelements in surface environments. As one looks to the future, the uncertainties multiply. The long-term effects of changes in land form and use, resulting from regional programmes such as the land reclamation projects of the Netherlands mentioned in Chapter 6, or a more ambitious proposal to ameliorate the climate of part of Siberia by diverting major rivers towards the Arctic Ocean, cannot be forecast with confidence. The development of nuclear power to fill the gap left by the depletion of fossil fuels has been slowed in many countries in the face of anxiety among the general public about the means of ensuring safe disposal of nuclear waste. More generally the probability that the climate will fluctuate in future as it has in the past has implications so far-reaching as to demand investigation. The rate and amount of the probable rise or fall of mean annual temperatures and the expectable changes of sea level resulting from waning or waxing of polar ice sheets can be guessed at by reference to the geological record of the Postglacial period (Figs. 9.4 & 5). The practical effects on water supply, on agriculture and on the stability of densely populated coastal zones, remain to be evaluated, as does the possible effect on climate of the release of CO_2 by combustion of coal and oil. In all these fields, the conventional distinction between pure and applied science is meaningless. Some of the responsibility for the decisions that will have to be made and acted upon rests on Earth scientists. It may be appropriate to end this book with a comment

Figure 9.5 Changes of sea level   The engraving of the Temple of Serapis, near Naples, used as the frontispiece of Charles Lyell's *Principles of geology*, illustrates submergence and partial emergence of the Italian coast since Roman times (see Section 6.4).

made by a contributor to a recent conference on atmospheric pollution:*

Our aim, in the end, is risk assessment. We shall never eliminate risk, but we should try to define more precisely the risks we are running, or asking others to run. We can decide in the end to accept the risks, but we ought to know what we are doing.

* Holdgate, M. W. 1979. *Phil Trans R. Soc. Lond.* **A290**, 607.

Appendix 1 Ore minerals

Ore minerals (Table A1.1) are defined by reference both to chemical composition and to the atomic lattice structure established by X-ray and other analytical techniques. Many minerals can, however, be identified beyond reasonable doubt in hand specimen or polished section by characteristic physical properties or by use of simple reactions diagnostic for the presence of a particular metal. The reader will be familiar with the use of crystallographic features and of physical properties in the identification of silicate minerals and the principles involved need no further elaboration (Table A1.2). The microscopic examination of ore minerals is, however, complicated by the fact that most sulphides, sulphosalts and metallic oxides are opaque and cannot be investigated by means of the standard petrological microscope.

Opaque minerals are studied in polished sections (mounted without a class cover) by means of a **polarising reflected light microscope**. A magnified image is obtained by means of a strong light beam directed vertically downwards onto the polished section and reflected up the microscope tube through a lens system to the eye. The insertion of polars in the paths of incident and reflected light makes it possible to establish the response to polarised light. The complexity of the optical system required to produce a reflected image and to offset the loss of light in the system makes the reflecting microscope a delicate and expensive instrument not always available for introductory courses.

The principal optical properties studied are colour, reflectivity and (for anisotropic minerals) bireflectivity or bireflexion. **Reflectivity** is a measure of the proportion of light reflected at the mineral surface and is apparent to the eye in the 'brightness' of the mineral. Relative reflectivities can be established by comparing minerals in a single section. Quantitative measurements range from below 10% for most silicates to over 90% for native silver.

A variety of other tests can be carried out on polished sections observed by means of the reflected light microscope. Relative hardness is indicated by resistance to polishing, hard minerals standing proud of softer neighbours. Quantitative estimates of **microhardness** are made by measuring the size of an indentation produced by a diamond point under a known weight. The ratio load : contact area is expressed on a numerical scale ranging up to about 2000. Among common minerals, galena, gold and molybdenite give values below 100, cassiterite, chromite and pyrite well over 1000. **Microchemical tests** involve bringing the section into contact with reagents for the suspected metal. Such tests may involve the use of impregnated paper which changes colour in the presence of a given metal when a specific reagent is added.

APPENDIX 1 ORE MINERALS

Table A1.1 Ore minerals: a basic list.

Metal	Mineral	Crystal system, habit	Density
aluminium	gibbsite $Al(OH)_3$		
	diaspore $AlO(OH)$		
antimony	stibnite Sb_2S_3	orthorhombic, (prismatic or bladed crystals)	4.5
arsenic	several arsenides		
chromium	chromite $(CrAlFe)_3O_4$	cubic (often octahedral)	4.5+
copper	native copper	cubic (sheets and filaments)	8.9
	chalcopyrite $CuFeS_2$	tetragonal (usually massive)	4.2
	bornite Cu_5FeS_4	cubic (usually massive)	5.0+
	chalcocite Cu_2S	orthorhombic (usually massive)	5.5+
	malachite $Cu_2(OH)_2CO_3$	monoclinic (usually botryoidal)	4.0
	cuprite Cu_2O	cubic (often octahedral)	5.8+
gold	native gold tellurides	cubic (scales, filaments, grains)	19
iron	magnetite Fe_3O_4	cubic (often octahedral)	5.1
	haematite Fe_2O_3	hexagonal (often scaly or massive)	4.9+
	goethite $Fe_2O_3H_2O$	orthorhombic (usually massive)	4.0+
	chamosite hydrated silicate	earthy	
	pyrite FeS_2	cubic (often striated cubes)	4.8+
	marcasite FeS_2	orthorhombic	4.9
	pyrrhotite $Fe_{1-x}S$	monoclinic (usually massive)	4.5
lead	galena PbS	cubic (often cubic form)	7.5
manganese	oxides, carbonates and silicates		
mercury	cinnabar HgS	hexagonal (usually massive)	8.1
molybdenum	molybdenite MoS_2	hexagonal (scales)	4.7
nickel	pentlandite $(FeNi)_9S_8$	cubic	5.0
	silicates of, e.g. olivine		
niobium	columbite (oxide)	orthorhombic	5.3+
platinum	alloys with Fe and other metals of Pt group		
silver	native silver substitution in PbZn sulphides	cubic (dendritic or massive)	10.1+
tin	cassiterite SnO	tetragonal (prismatic)	6.8+
titanium	ilmenite $FeTiO_3$	hexagonal	4.5+
tungsten	wolframite $(FeMn)WO_4$	monoclinic (tabular)	7.1+
	scheelite $CaWO_4$	tetragonal (usually massive)	5.9+
uranium	uraninite UO_2	cubic (usually massive)	8.0
	pitchblende UO_2	amorphous	
zinc	sphalerite ZnS	cubic	3.9+
	smithsonite (carbonate)	hexagonal	4.0+
	hemimorphite (silicate)	orthorhombic	3.4

The symbol + under relative density indicates a range of values upward from the figure given.

Table A1.2   Diagnostic properties used in identification of ore minerals.

Physical properties		*Physical properties*	
crystallographic common habit	crystal system, symmetry, common forms, twinning etc.	fluorescence	scheelite and a few other minerals fluoresce on exposure to ultraviolet light
colour, streak, lustre	most native metals, sulphides and sulphosalts have metallic lustre	radioactivity	emission of particles by U, Th recorded by geiger counter, discolours sensitive film
relative density	most ore minerals are more dense than most silicates, Au, Ag and minerals containing Pb and Au have very high densities	optical properties	reflectivity and bireflectance
		Chemical properties	
		alteration products	oxidation of sulphides and sulphosalts containing certain metals produces distinctive alteration products (e.g. peacock-blue staining with copper, grey frond-like growths with manganese)
hardness	specimens classified in terms of Mohs' scale; microhardness by Vickers hardness number (soft <85, medium 85–230, hard >230)		
magnetic properties	magnetite strongly magnetic, pyrrhotite more weakly magnetic: some other minerals are attracted by the electromagnet	microchemical tests	applied to polished sections, e.g. contact print method, impregnated paper initiates specific reactions signified by colour change
electrical conductivity	high in native metals, most sulphides and sulphosalts	full analysis	usually by electron microprobe
		lattice structure	X-ray diffraction methods

Index

References to figure numbers appear in *italics*, and references to tables appear in **bold type**.